TRANSMISSION LINE MATRIX (TLM) TECHNIQUES FOR DIFFUSION APPLICATIONS

TRANSMISSION LINE MATRIX (TLM) TECHNIQUES FOR DIFFUSION APPLICATIONS

Donard de Cogan

Reader in Electronics
School of Information Systems
University of East Anglia, Norwich

CRC Press is an imprint of the
Taylor & Francis Group, an **informa** business

Amsteldijk 166
1st Floor
1079 LH Amsterdam
The Netherlands

Cover design: diffusion plot created by K. L. Moravec

British Library Cataloguing in Publication Data

De Cogan, Donard
Transmission line matrix techniques for diffusion applications
1. Microwave transmission lines – Mathematical models
I. Title
621.3

ISBN 9056991299

A catalogue record for this book is available from the British Library.

Contents

Preface

It is quite unusual for a scientist or engineer to be able to say "I was there when it started." I was not there when Transmission Line Matrix (TLM) was born, but I can say that I watched it develop from infancy.

I had spent three years making and evaluating baritt (punch-through) microwave diodes and had been endeavouring to find some uses for those that would not oscillate. The invariance of voltage over a significant temperature range was a discovery, while use as transient suppressers seemed tediously obvious. Appearances can be deceptive and the obvious is frequently more complex. While working at Mullard power semiconductors I fabricated large area punch-through devices which were then tested. However, the equipment was not quite suitable. I felt that a transient suppresser should be tested for its intrinsic properties, not the quality of die bonding, packaging and heat-sinking. Thus the construction of fast transient test equipment was one of my first objectives when I started as a lecturer in electrical and electronic engineering at Nottingham University. The results of measurements were strange; simulations were needed, but how should I set about it?

Finite difference methods would have been the first choice, but my colleague, Peter Johns, said "why not try my TLM technique?". He was like a child with a new toy that could do everying. In fact, he was a pioneer in the 'new engineering'. Generations of students had been turned away from the subject by endless hours of seemingly irrelevant mathematics. They first acquired the tools (techniques for treating various differential equations) and then addressed the initial and boundary conditions. All of this was followed by the realisation that such analysis rarely matched reality without the excessive use of assumptions; indeed part of the training as an engineer was to learn to recognise which assumptions retained the essence of the original problem.

Johns' view was that an electrical network with transmission lines could yield an accurate analogue description of a physical problem. Thus the solution of the currents and voltages within the network matrix was in fact the solution of the problem. There was no need to progress via a mathematical description. He argued that the construction of the transmission line matrix was intuitive for the engineer.

Although Johns had demonstrated that TLM could be used for diffusion problems, most of his energies were taken up with extending the technique for electromagnetic applications. Lossless TLM had first been demonstrated in two dimensions: the determination of the resonances in a lossless waveguide. The extension to three dimensions was not straightforward, but I will always remember his excitement when he said "... I was shaving this morning when it came to me. It is so elegant." That was the start of the symmetrical condensed node which is now a cornerstone in lossless TLM.

It is sometimes difficult to be critical of a hero, but let me say that I frequently encountered difficulties in understanding Johns' explanations of the technique. The ideas were sublime, but catching the detail was the problem, so I had to get on with developing my diffusion algorithms by trial and much error. There was no appreciation of the differences between link-line and link-resistor nodes, nor of the fact that current (equivalent of matter flow) had to be conserved. How was the capacitance to be distributed in thermal models, and was it R or 2R? Nevertheless, Johns was patient and when we did simulate semiconductor carrier recombination by means of link transmission lines which had an exponential decay of carriers along their length, he declared this approach to be elegant, a perfectly intuitive application of physical concepts.

Before his death Johns worried much about whether TLM would ever become accepted. He saw finite difference and finite element methods as the giants which would swallow up his baby. He held the view that any advantages that TLM had at that time would be reduced as computing power increased. He failed to realise that the same computing power would create new vistas for TLM. Nevertheless, he did live to see the first stages in the diffusion of his gospel. Wolfgang Hoefer's group as well as others in Canada and elsewhere took up the torch and the *International Journal of Numerical Modelling* was started. Researchers world-wide have since extended our knowledge of the technique itself and Christos Christopoulos has written the book dedicated to TLM.

In the meantime diffusion TLM has moved forward, although not at quite the same pace. Johns obtained several research contracts to do work on heat-flow in jet engines and on diffusion processes in food. He was ably assisted by Susan Pulko who established her own research group when she moved to Hull University. My early progress was assisted by Susan John, Mohamed Henini and Achal Shah. The collaborations with the late Adnan Saleh and with Peter Enders have been particularly productive. I will always remember the frisson between Adnan, who solved everything with an electrical network, and Peter, whose approach to any problem was to start with a Taylor expansion. Xiang Gui came to UEA for two weeks to learn about TLM. We wrote a paper before he left and this laid the foundations for our continuing collaboration.

There were many reasons for writing this book. Perhaps the hardest thing about TLM is the initial barrier of the mind. In addition to the diverse concepts on which it draws, there is also the representational language. I have been conscious that as the discipline expands, so there is the opportunity for different dialects to develop. Although there is nothing intrinsically wrong with this, it could inhibit rapid communication across the research community. Thus it is hoped that this book will provide a tutorial guide and will act as a focus for a common terminology which is accessible by a wide audience.

There is also some evidence of the beginnings of priority disputes (who was it who developed a particular technique/algorithm/application first?). This problem could increase as the subject expands and so I have attempted to review the current state of research in each area of application and to present the developments in a historical sequence.

Individual acknowledgements are difficult because I should really include everyone who has contributed to TLM. Perhaps my closing acknowledgement should be to my family who have tolerated my obsession and my 0530 starts for so long.

Chapter 1

Introduction

While working at EEV Chelmsford Raymond Beurle identified a specific need to express electromagnetic phenomena in the time domain. He had used an early computer to simulate the propagation of activity in neural networks and later had experience of using Southwell's relaxation technique to solve electrostatic field problems. These two apparently unrelated themes coalesced to suggest that propagation in a matrix of transmission lines might be used to simulate propagation in space, to enable high frequency field distributions to be calculated in arbitrarily shaped cavities.

A TLM matrix was deliberately chosen in preference to a network of finite L and C elements because it so greatly simplified the theory regarding the interaction between a short pulse of voltage (or current) and each mode. Another advantage was that a finite amount of energy introduced at a source in the matrix could not increase and the calculation was therefore unconditionally stable, thus avoiding a problem that has been encountered with some other methods of calculation.

After a small trial confirmed propagation and reflection at a boundary in TLM the ideas was suggested to a postgraduate student who subsequently reported confirming this with a computer simulation. Some time later Peter Johns, a microwave engineer at the Post Office Telecommunications Research laboratory at Dollis Hill, London was appointed as a lecturer at Nottingham. He asked Beurle to suggest a research topic, and as no mathematically minded postgraduate student had come forward to take this topic Beurle felt (rightly as it transpired) that this would be a good way of launching TLM. Events proved that this was indeed so. Peter Johns took up the idea with an enthusiasm that has become legendary. The details of the method were first published in 1971 [1.1] and Beurle was asked to co-author this first paper as an acknowledgement of the source of the idea.

The approach is based entirely upon establishing an analogue between a space and time dependent physical problem and an electrical network. This in itself has a long tradition in modelling and simulation. The novelty of TLM lay in the inclusion of lengths of transmission line which imposed time-delay. Adaptations quickly followed [1.2 – 1.4] and the wider potential was soon recognised amongst a small but dedicated group of researchers.

Over the next few years developments in TLM were largely concerned with electromagnetic applications, although the potential for diffusion modelling had been outlined [1.5]. In 1978 the author joined the staff of Nottingham University and was first involved in experimental research on the impulse withstand capability of electrical transient suppressors. He was strongly encouraged by Johns to use TLM to model the processes of thermal overload. The successes in this area immediately led to wider interest, the investigation of wider applications and an exponential growth in the number of publications.

At first there were only a small number of research groups working on lossy TLM. Some were concerned mainly with theory while others were concerned with applications, albeit within specific fields. The situation is now changing rapidly. There are now many who have heard about TLM, who would like to know what it is, what it can do and how to get started.

The book is divided into two parts; theory and applications. Although the main interest is in diffusion-related techniques we adopt the conventional approach and start with a chapter on lossless TLM. Here we present an outline of the theory and some relevant algorithms. The author subscribes to the school of thought which believes that where possible *how does it work*? comes before the heavy details of *why does it work*? For that reason techniques and theory are covered in chapters 3 and 4 respectively.

Within the applications part of the book there has been an attempt to follow a historical sequence: heat-flow comes first, then matter diffusion and then all the other ingenious applications which makes this such a rewarding field of research.

It is useful, but not essential for the reader to assimilate the background electromagnetic theory. Indeed it is quite possible to adopt the algorithms as a starting point and to use them. However, it is felt that this looses much of the point of TLM modelling; an application-specific nodal or network arrangement which is consistent with electromagnetic principles is unlikely to lead to a paradoxical outcome.

Acknowledgements

All who knew Peter Johns acknowledge his unique contribution. The breadth of topics which are presented in this book could hardly have been conceived without his initial insight. I have been particularly fortunate to have the continuing encouragement of Raymond Beurle and of friends and colleagues such as Peter Enders, Xiang Gui, Michael Malachowski, Susan Pulko, Paul Webb, Tony Wilkinson and many others. I would also like to acknowledge the contribution of colleagues such as Stephen Cox and ex-students: Tim Lee, Kimberley Moravec and Alison Smith whose collaborations may never have seen the light of publication, but which nevertheless add significantly to topics which are covered in this book.

References

[1.1] P. B. Johns and R. L. Beurle, Numerical solution of 2-dimensional scattering problems using a transmission line matrix, *Proceedings of the IEE*, 188 (1971), 1203 – 1208.

[1.2] P. B. Johns, Application of the transmission-line matrix method to homogeneous waveguides of arbitrary cross-section, *Proceedings IEE*, **119** (1972), 1986 – 1991.

[1.3] P. B. Johns and G. F. Slater, Transient analysis of waveguides with curved boundaries, *Electronic letters*, 9 (1973), 486 – 487.

[1.4] P. B. Johns, The solution of inhomogeneous waveguide problems using a transmission-line matrix, *IEEE Trans. Microwave Theory and Techniques*, **MTT-22** (1974), 209 – 215.

[1.5] P. B. Johns, A simple, explicit and unconditionally stable routine for the solution of the diffusion equation, *Int. Jnl. Numerical Methods in Engineering*, **11** (1977), 1307 – 1328.

Chapter 2

The theory and implementation of lossless TLM algorithms

This chapter outlines the evolution of lossless TLM algorithms. In order to present a firm foundation we have to talk about propagation and about transmission lines themselves. We start by presenting some of the underlying electromagnetic theory which is relevant to propagation in homogeneous free-space. This is followed by a brief discussion on the use of analogues in electromagnetics. We then move on to compare lumped and distributed circuits which is the starting point for an introduction to the theory of transmission lines.

Propagation of a signal in loss-free space

Maxwell's equations for the propagation of a transverse magnetic wave in two dimensions can be expressed as:

$$\frac{\partial E_z}{\partial x} = -\mu\,\frac{\partial H_y}{\partial t} \qquad \frac{\partial E_z}{\partial y} = -\mu\,\frac{\partial H_x}{\partial t} \tag{2.1}$$

$$\frac{\partial H_y}{\partial x} + \frac{\partial H_x}{\partial y} = \sigma E_z + \varepsilon\frac{\partial E_z}{\partial t} \tag{2.2}$$

The electric and magnetic field components which contribute to these are shown in figure 2.1.

Manipulation of these equations leads to

$$\frac{\partial^2 E_z}{\partial x^2} + \frac{\partial^2 E_z}{\partial y^2} = \mu\varepsilon \frac{\partial^2 E_z}{\partial t^2} + \varepsilon\sigma \frac{\partial E_z}{\partial t} \tag{2.3}$$

The term containing $\mu\sigma$ expresses a wave propagation with a velocity $\mu\varepsilon = 1/c^2$. The second term, which contains $\varepsilon\sigma$ expresses an attenuation of the wave.

Electromagnetic analogues

For reasons of conceptual visualisation or efficiency of calculation it is often convenient to describe a problem by means of one or more analogues which can then be translated into a mathematical description. The use of electrical networks for this purpose has a long tradition. A mechanical or hydraulic

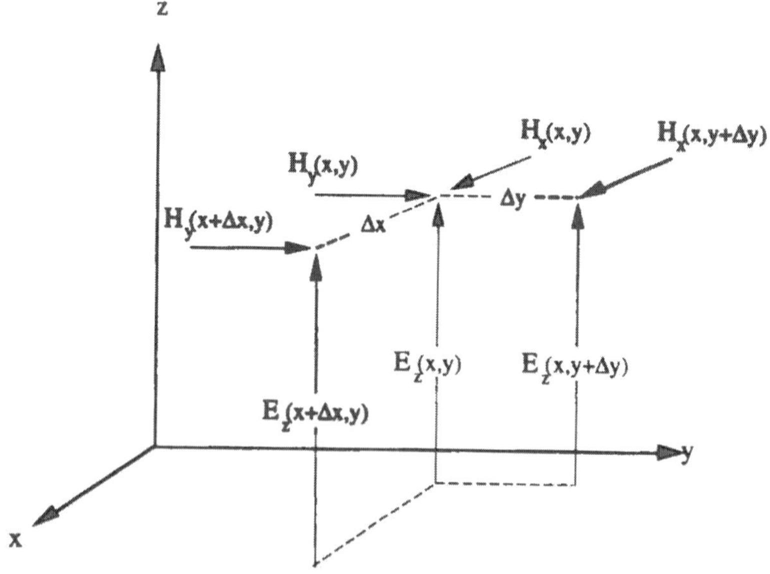

Figure 2.1 The spatial variation of the electric and magnetic field components which contribute to eqns (2.1) and (2.2).

system can be replaced by an electrical network. This is then activated by the application of currents/voltages and the subsequent response provides a direct solution of the original physical problem. The simulation of fluid-flow in a pipe due to pressure difference by using current and voltage analogues is an example of the application of Ohm's law.

An identical approach can also be used to obtain analogue solutions of electromagnetic problems. The fields around some complex geometry such as the rotor/stator in an electric motor can be estimated in two dimensions by placing metal cross-sections on a sheet of conductive paper. If a DC source is connected between the conductors then current will spread out across the two-dimensional resistance of the paper and will flow along the field lines according to Ohm's law. A potentiometer can then be used to locate the lines of equal potential. Free-hand drawing of curvilinear squares yields the density of lines which are orthogonal to the equipotentials (the electric field) [2.1].

Kron, whose work provided much of the inspiration for TLM, proposed electrical network analogues for the electromagnetic equations [2.2], [2.3]. Before the advent of computers physical networks were constructed and appropriate measurements provided the required result [2.4]. More recently, Hammond and Sykulski [2.5] have developed analogue techniques in which potential represents the electric field and current represents the magnetic field.

Distributed and lumped circuits

The ideas which have been outlined above can also be expressed for a lumped component electrical network consisting of resistors, capacitors and inductors. In this case we will use the analogues $V \equiv E$ and $I \equiv H$. We will start by considering a one-dimensional case (figure 2.2) which has lumped components whose values are equivalent to what are called distributed parameters. Thus, R_d is the resistance per unit length $= R/\Delta x$. C_d and L_d are similarly defined.

The change in voltage as a function of distance depends on the voltage drop across the resistor (IR) and the emf across the inductor ($-LdI/dt$).

Thus

$$\lim \Delta x \to 0 \frac{V + \Delta V - V}{\Delta x} = \frac{\partial V}{\partial x} = -IR_d - L_d \frac{\partial I}{\partial t} \qquad (2.4)$$

At the same time there is a change in current over this distance due to the change in charge being stored on the capacitor.

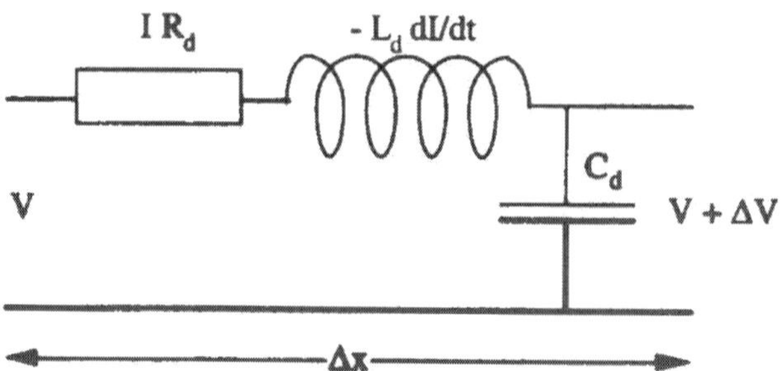

Figure 2.2 A simple LRC circuit.

$$\begin{aligned} I = C\frac{\partial V}{\partial t} &= C_d \Delta x \, \frac{\partial V}{\partial t} \text{ so that} \\ \lim \Delta x \to 0 \frac{I + \Delta I - I}{\Delta x} &= \frac{\partial I}{\partial x} = -\, C_d \, \frac{\partial V}{\partial t} \end{aligned} \tag{2.5}$$

Now, if eqn (2.4) is differentiated with respect to x then we get

$$\frac{\partial^2 V}{\partial x^2} = -\, R_d \, \frac{\partial I}{\partial x} - L_d \, \frac{\partial^2 I}{\partial x \partial t} \tag{2.6}$$

Eqn (2.5) and its derivative can now be substituted into this to give:

$$\frac{\partial^2 V}{\partial x^2} = L_d C_d \, \frac{\partial V^2}{\partial t^2} + R_d C_d \, \frac{\partial V}{\partial t} \tag{2.7}$$

This is called the *Telegraphers'* equation and is the basis for the TLM method. The extension of this equation to two and three dimensions will be deferred until we consider specific TLM networks. However, before we proceed further it is necessary to cover some additional basic electromagnetic theory.

Transmission lines

The concept of distributed parameters which was mentioned is central to our model for a transmission line. This is an entity which has capacitance, inductance and resistance distributed along its length[1]. The electromagnetic

[1]Strictly speaking, there will be time delay in any real electrical component, but for normal analysis devices such as resistors, capacitors and inductors are treated as ideal so that the inherent delay is ignored and only the delay in transmision lines is considered.

behaviour can be expressed using eqns (2.1–2.3). So long as R_d is negligible then we can easily define some important parameters:

the impedance

$$Z = \sqrt{\frac{L_d}{C_d}} = \sqrt{\frac{L}{C}} \tag{2.8a}$$

the velocity of an impulse on a line

$$v = \sqrt{\frac{1}{L_d C_d}} \tag{2.8b}$$

the propagation constant, γ of an impulse on a line

$$\gamma = j\omega\sqrt{L_d C_d} \tag{2.8c}$$

The time taken for an electrical signal impressed on the line to traverse it will be determined by the local velocity of light, which depends on $\mu\epsilon$ as mentioned above for free space.

If R_d is non-negligible then the situation is much more complicated

$$Z = \sqrt{\frac{R + j\omega L}{j\omega C}} \tag{2.9}$$

$$\gamma = \sqrt{(R + j\omega L)j\omega C} \tag{2.10}$$

For many years the textbook approaches to electromagnetics have explored ways of circumventing this problem. If R_d is negligible then the Telegraphers' equation reduces to a simple wave equation and this is the basis for lossless TLM modelling. If on the other hand R_d is significant then there is an entire branch of TLM modelling which seeks to ignore the wave component in eqn (2.7) so that it can be treated as a diffusion equation. Before addressing this, which after all is the main subject matter of this book we will look at the development of lossless TLM algorithms in one and two-dimensions.

The velocity of propagation on a uniform, lossless transmission line can be related to its parameters:

$$v = \frac{\Delta x}{\Delta t} = \sqrt{\frac{1}{L_d C_d}} = \sqrt{\frac{\Delta x^2}{LC}} = \Delta x\sqrt{\frac{1}{LC}} \tag{2.11}$$

or

$$\frac{1}{\Delta t} = \frac{1}{\sqrt{LC}} \tag{2.12}$$

now, using $Z = \sqrt{L/C}$ we can eliminate either L or C in the eqn (2.12) to get

$$Z = \frac{\Delta t}{C} \quad \text{or} \quad \frac{\Delta t}{C_d \Delta x} \tag{2.13}$$

$$Z = \frac{L}{\Delta t} \quad \text{or} \quad \frac{L_d \Delta_x}{\Delta t} \tag{2.14}$$

These relationships between the line impedance, the line parameters and the spatial and temporal discretisations are the fundamental building blocks of TLM.

Reflections at impedance discontinuities in transmission lines

If a signal is travelling along a lossless transmission line then it will continue unperturbed provided that the impedance of the line, Z_0 remains unchanged. If however, the impedance of the line at some point becomes Z_T then a portion of the signal will be reflected back on itself. The reflection coefficient is then given by the equation:

$$\rho = \frac{Z_T - Z_0}{Z_T + Z_0} \tag{2.15}$$

The signal which is transmitted past a discontinuity in a line depends on whether we are dealing with current or voltage. If we are dealing with current (charge per second) then the conservation of matter controls what happens:

$${}^{i}I = {}^{r}I + {}^{t}I \tag{2.16}$$

in this case the super-scripts i, r and t indicate whether the current is *incident, reflected* or *transmitted.* The reflected current can be defined using the reflection coefficient as:

$${}^{r}I = \rho\, {}^{i}I \tag{2.17}$$

therefore the transmitted current is:

$${}^{t}I = (1 - \rho)\, {}^{i}I \tag{2.18}$$

An analysis of the scattering of voltage pulses (figure 2.3) starts from the same position but recalls that voltage is a measure of work done when charge is moved against an electric field. Equation (2.16) can be restated using Ohm's law as:

$$\frac{{}^tV}{Z_0} = \rho \frac{{}^iV}{Z_0} + \frac{{}^tV}{Z_T} \qquad (2.19)$$

However, by virtue of eqn (2.15) we can write this in terms of ${}^tV$ as:

$${}^tV = (1-\rho)\,{}^iV \frac{Z_T}{Z_0} = (1+\rho)\,{}^iV \qquad (2.20)$$

Both the current and voltage expressions for transmission and reflection are essential in the development of lossless and lossy TLM algorithms and a consequence of an early misinterpretation of these features is discussed in chapter 5.

TLM nodal configurations

Conventional TLM has used what is called the *shunt* node configuration. This assumes that current is equivalent to magnetic field and voltage is equivalent to electric field. In a two dimensional problem I_x and I_y are the analogue of H_x and H_y and V_z is equivalent to E_z. There is another approach called the *series* node configuration which uses E_x, E_y (equivalent to V_x, V_y) and H_z equivalent to I_z. Details of the implementation of this and the more complicated three-dimensional lossless representations can be found in references such as Christopoulos [2.6]. We will concentrate here on the development of lossless TLM algorithms based on the shunt node in one and two dimensions.

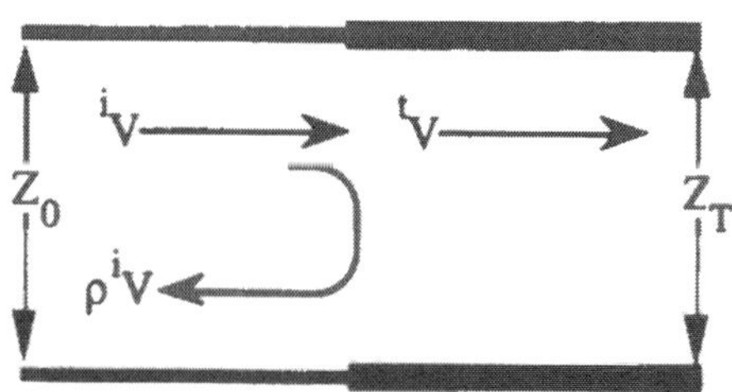

Figure 2.3 Voltage scattering at an impedance discontinuity.

Lossless shunt mesh TLM algorithms

Figure 2.4(a) shows a single two dimensional cell formed of transmission lines, where each arm links to other similar cells. The entire space of the problem is filled out as a Cartesian mesh of these cells. A lumped circuit representation can be seen in figure 2.4(b). The inductors are series elements and the capacitor is a *shunt* element, hence the nomenclature. In terms of the field components in Maxwell's electromagnetic equations this network yields H_x, H_y and E_z. By analogy these can be directly related to currents and voltage in a circuit.

The conversion of these ideas into practical TLM algorithms requires the use of two simple assumptions, one equation from electromagnetics and one theorem from basic electrical theory:

- The first assumption is that all data (field amplitudes) are represented by impulses of very short duration. Thus an impulse entering a transmission line has no knowledge about the length of the line; indeed it is unaware that the line is finite until it arrives at a discontinuity.
- The second assumption is needed for computational purposes and requires that all pulses move around the spatial mesh in synchronism.
- We also assume that the electromagnetic equation for reflection at an impedance discontinuity (Eqn (2.15)) applies to these pulses.
- The final item that is required is Thévenin's theorem which describes the circuit equivalent of a transmission line in terms of the open-circuit voltage and the short-circuit impedance. The impedance is an easy matter: if we look in at one end of a transmission line with the other end shorted, then the measured impedance is Z. In order to visualise the open

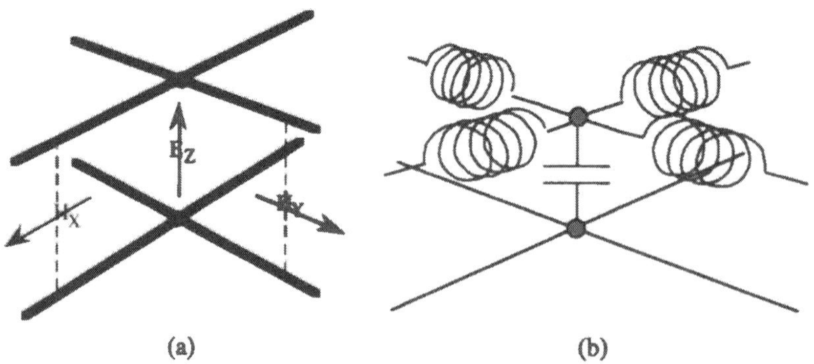

(a) (b)

Figure 2.4 (a) intersecting pair of two-wire lossless transmission lines (b) lumped circuit equivalent for (a).

circuit voltage, we must consider an incoming impulse of magnitude ${}^iV$ as it approaches the observation point. By the Thévenin definition this is an open circuit ($Z_T = \infty$) and thus $\rho = 1$. Therefore, at the instant of arrival we have the superposition of the incident pulse (${}^iV$) and the reflected pulse (${}^rV = {}^iV$). Accordingly, the transmission line can be represented by an impedance Z in series with a voltage source of magnitude $2\,{}^iV$.

We are now in a position to monitor the progress of a single pulse as it starts to scatter around the mesh. Since we are dealing with a Cartesian mesh, we can describe the directions of scatter by the compass directions N, S, E and W. Let us consider ${}^iV_W$, which is incident from the west onto a two-dimensional TLM node (x,y) as shown on the left in figure 2.5.

As the pulse arrives at the end of its line what it sees in front of it and how it appears are shown on the right above. Thus the pulse will undergo scattering according to eqn (2.15). In this case Z_T is equal to $Z/3$, the three impedances seen in parallel by the incoming pulse, and thus $\rho = -1/2$. This means that a pulse of magnitude $-{}^iV_W/2$ is returned down the incoming transmission line. The remainder of the signal is transmitted into the other arms. Pulses incident from arms N, S and E are simultaneously incident and undergo scattering. They also contribute to the voltage at the node centre which can be represented by the superposition of all voltages, giving the overall voltage at the node at the k-th time interval:

$$ {}_k\phi(x,y) = \frac{1}{2}({}^iV_N + {}^iV_S + {}^iV_E + {}^iV_W) \tag{2.21} $$

The pulse which is scattered back in any direction is the sum of what is reflected and what is transmitted from all other arms. Thus

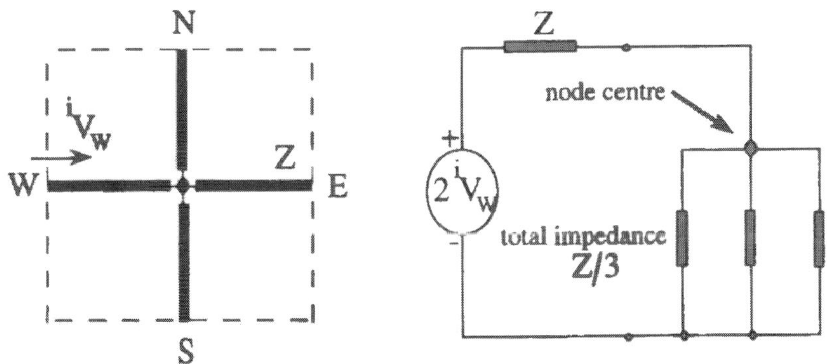

Figure 2.5 Lossless TLM node and its Th venin equivalent.

$$ {}^{s}V_{\mathrm{W}} = \rho\, {}^{i}V_{\mathrm{W}} + \tau({}^{i}V_{\mathrm{N}} + {}^{i}V_{\mathrm{S}} + {}^{i}V_{\mathrm{E}}) \tag{2.22} $$

There are similar equations for ${}^{s}V_{\mathrm{N}}$, ${}^{s}V_{\mathrm{S}}$, ${}^{s}V_{\mathrm{E}}$ and ${}^{s}V_{\mathrm{W}}$ and the entire scattering process can be expressed in matrix form:

$$ {}_{k}\begin{pmatrix} {}^{s}V_{\mathrm{N}} \\ {}^{s}V_{\mathrm{S}} \\ {}^{s}V_{\mathrm{E}} \\ {}^{s}V_{\mathrm{W}} \end{pmatrix} = \frac{1}{2}\begin{pmatrix} -1 & 1 & 1 & 1 \\ 1 & -1 & 1 & 1 \\ 1 & 1 & -1 & 1 \\ 1 & 1 & 1 & -1 \end{pmatrix} {}_{k}\begin{pmatrix} {}^{i}V_{\mathrm{N}} \\ {}^{i}V_{\mathrm{S}} \\ {}^{i}V_{\mathrm{E}} \\ {}^{i}V_{\mathrm{W}} \end{pmatrix} \tag{2.23} $$

Each pulse travels the discretisation distance Δx during the discretisation time Δt after which it becomes an incident pulse at an adjacent node. The connections to other nodes as seen at node (x,y) can be expressed in terms of space and time-step, $k+1$ as

$$ \begin{aligned} {}_{k+1}^{\;\;i}V_{\mathrm{N}}(x,y) &= {}_{k}^{s}V_{\mathrm{S}}(x, y+1) \\ {}_{k+1}^{\;\;i}V_{\mathrm{S}}(x,y) &= {}_{k}^{s}V_{\mathrm{N}}(x, y-1) \\ {}_{k+1}^{\;\;i}V_{\mathrm{E}}(x,y) &= {}_{k}^{s}V_{\mathrm{W}}(x+1, y) \\ {}_{k+1}^{\;\;i}V_{\mathrm{W}}(x,y) &= {}_{k}^{s}V_{\mathrm{E}}(x-1, y) \end{aligned} \tag{2.24} $$

The repeated application of eqns. (2.21), (2.23) and (2.24) for every time step allows us to describe the propagation of the single initial impulse throughout the mesh as a function of time.

The *True Basic* code for two-dimensional lossless propagation which is given below considers a group of four pulses, each of magnitude 100, which are injected into point (10,10) within a 20×20 mesh at time $t=0$. The programme prints out the value of $\phi(10,10)$ (designated VNODE(10,10)) and its nearest neighbours in the x-direction during 5 iterations. The seemingly peculiar ranges of i and j in the connect subroutines are intended to avoid any of the values falling outside the bounds of the array size definitions. In most practical examples we are either dealing with very large arrays and/or have clearly defined boundary conditions which avoid these problems. The question of boundaries will be addressed later in this chapter.

```
REM - -TLM programme - -
DIM VSN(20,20),VSS(20,20),VSE(20,20),VSW(20,20)
DIM VIN(20,20),VIS(20,20),VIE(20,20),VIW(20,20)
DIM VNODE(20,20)

REM - - - - - - - - - - INPUT - - - - - - - - - - -
  LET VIN(10,10) = 100
  LET VIS(10,10) = 100
```

```
  LET VIE(10,10) = 100
  LET VIW(10,10) = 100

REM - - - - - - - - - - main - - - - - - - - - - -
  FOR K = 1 TO 5
   CALL tlm
   CALL printout
  NEXT K

REM - - - - - - - - - end of main - - - - - - - - -

REM - - - - - - - - - THE SUBROUTINES - - - - - - - - -
SUB tlm
  CALL summation
  CALL scatter
  CALL connect
END SUB

REM - - - - - - - - - - - - - - - - - - - - - - - - - - - - - - - - - - - - - - - - -
SUB summation
  FOR I = 1 TO 20
    FOR J = 1 TO 20
      LET VNODE(I,J) = 0.5 * (VIN(I,J)+VIS(I,J)+VIE(I,J)+VIW(I,J))
    NEXT J
  NEXT I
END SUB

REM - - - - - - - - - - - - - - - - - - - - - - - - - - - - - - - - - - - - - - - -
SUB scatter
  FOR I = 1 TO 20
    FOR J = 1 TO 20
      LET VSN(I,J)=-0.5*VIN(I,J)+0.5*(VIS(I,J)+VIE(I,J)+VIW(I,J))
      LET VSS(I,J) =-0.5*VIS(I,J)+0.5*(VIN(I,J)+VIE(I,J)+VIW(I,J))
      LET VSE(I,J)=-0.5*VIE(I,J)+0.5*(VIS(I,J)+VIN(I,J)+VIW(I,J))
      LET VSW(I,J)=-0.5*VIW(I,J)+0.5*(VIS(I,J)+VIE(I,J)+VIN(I,J))
    NEXT J
  NEXT I
END SUB

REM - - - - - - - - - - - - - - - - - - - - - - - - - - - - - - - - - - - - - - - -
SUB connect

  REM *THE CONNECT PROCESSES IN THE Y-DIRECTION*
    FOR I = 1 TO 20
      FOR J = 2 TO 19
        LET VIN(I,J) = VSS(I,J+1)
```

```
          LET VIS(I,J) = VSN(I,J-1)
        NEXT J
      NEXT I

REM *THE CONNECT PROCESSES IN THE X-DIRECTION*
  FOR I = 2 TO 19
    FOR J = 1 TO 20
      LET VIE(I,J) = VSW(I+1,J)
      LET VIW(I,J) = VSE(I-1,J)
    NEXT J
  NEXT I
END SUB
REM ----------------------------------------
SUB printout
  PRINT " "
  PRINT K,VNODE(9,10), VNODE(10,10), VNODE(11,10)
END SUB

END
```

TLM algorithms can also be implemented using matrix transformations which is particularly easy in MATLAB. Hansleman [2.7] provides a collection of *m*-files for moving the contents of a matrix either up, down left or right. These are essentially Toeplitz matrices, but we will call them 'shift matrices'. If we start by considering a matrix with a single non-zero value then we can investigate the effect of various shift matrices:

$$\mathbf{M} = \begin{bmatrix} 0 & 0 & 0 & 0 & 0 \\ 0 & 0 & 0 & 0 & 0 \\ 0 & 0 & 1 & 0 & 0 \\ 0 & 0 & 0 & 0 & 0 \\ 0 & 0 & 0 & 0 & 0 \end{bmatrix}$$

We consider two shift matrices:

$$\mathbf{S}_1 = \begin{bmatrix} 0 & 1 & 0 & 0 & 0 \\ 0 & 0 & 1 & 0 & 0 \\ 0 & 0 & 0 & 1 & 0 \\ 0 & 0 & 0 & 0 & 1 \\ 0 & 0 & 0 & 0 & 0 \end{bmatrix} \quad \text{and} \quad \mathbf{S}_2 = \begin{bmatrix} 0 & 0 & 0 & 0 & 0 \\ 1 & 0 & 0 & 0 & 0 \\ 0 & 1 & 0 & 0 & 0 \\ 0 & 0 & 1 & 0 & 0 \\ 0 & 0 & 0 & 1 & 0 \end{bmatrix}$$

It can be shown that the matrix product $\mathbf{S}_1\mathbf{M}$ is equivalent to a *shift-up* transformation, while $\mathbf{S}_2\mathbf{M}$ is equivalent to a *shift-down* transformation. The products $\mathbf{MS}_1$ and $\mathbf{MS}_2$ shift the contents of **M** to left and right respectively.

Matrix scatter and connect processes of TLM can be joined together as:

$$\begin{aligned}
{}_{k+1}^{\ i}\mathbf{V}_N(x,y) &= [\text{shift-up}][\rho\,{}_k^i\mathbf{V}_N + \tau\,{}_k^i\mathbf{V}_S + \tau\,{}_k^i\mathbf{V}_E + \tau\,{}_k^i\mathbf{V}_W] \\
{}_{k+1}^{\ i}\mathbf{V}_S(x,y) &= [\text{shift-down}][\tau\,{}_k^i\mathbf{V}_N + \rho\,{}_k^i\mathbf{V}_S + \tau\,{}_k^i\mathbf{V}_E + \tau\,{}_k^i\mathbf{V}_W] \\
{}_{k+1}^{\ i}\mathbf{V}_E(x,y) &= [\text{shift-left}][\tau\,{}_k^i\mathbf{V}_N + \tau\,{}_k^i\mathbf{V}_S + \rho\,{}_k^i\mathbf{V}_E + \tau\,{}_k^i\mathbf{V}_W] \\
{}_{k+1}^{\ i}\mathbf{V}_W(x,y) &= [\text{shift-right}][\tau\,{}_k^i\mathbf{V}_N + \tau\,{}_k^i\mathbf{V}_S + \tau\,{}_k^i\mathbf{V}_E + \rho\,{}_k^i\mathbf{V}_W]
\end{aligned} \tag{2.24}$$

In this case ρ is -0.5 and τ is 0.5, but the algorithm can be applied to more general situations. Additionally, if it is implemented using a symbolic package then it is possible to see the propagation process in an algebraic way.

Boundaries

Naturally, we cannot have an infinitely large mesh and the question of boundaries therefore arises. TLM starts by using the same boundaries as would be used in the mathematical analysis of microwaves.

- If there is an open-circuit termination, then $Z_T = \infty$. This means that pulses which arrive at a boundary are reflected in-phase, because $\rho = 1$.
- If on the other hand there is a short-circuit then $Z_T = 0$. The reflection coefficient, $\rho = -1$ and any pulse incident on a boundary will be reflected in anti-phase.
- In the situation where $Z_T = Z$ then $\rho = 0$. This is called a 'matched load' boundary condition and will feature in later discussions.

The concept of open-circuit boundaries can also be used to reduce the size of the problem which needs to be computed. $\rho = 1$ does not distinguish between a pulse which is reflected in phase at a boundary and one which passes through a symmetry axis. Consequently, in the case of the rectangular wave-guide that was analysed by Johns and Beurle [1.1] the entire cross-section of the waveguide could be analysed by having two short-circuit boundaries (the outer walls and two open circuit boundaries, the two orthogonal symmetry axes.

Some of the above ideas can be demonstrated by means of the MATLAB programme which is shown below. This concerns the one-dimensional propagation of an input in a 20 node mesh which $\rho = -1$ (short-circuit) boundaries at both extremities. The input is distributed along the mesh and consists of three sinusoidal waves with frequencies ω, 3ω and 4ω with relative amplitudes 0.5, 0.1 and 0.05. At $t = 0$ this is described by equal opposite excitations at each node (vir($x(j)$) and vil($x(j)$)). Hansleman's [2.7]

Shift routines are used to achieve the *Connect* processes. It will be seen that setting the values of vil(1) = −vsl(1) and vir(20) = −vsr(20) achieves the correct description of the two boundaries. In this example the process is allowed to continue for 512 iterations. The instantaneous value of the amplitude at node 10 is recorded at every iteration and is stored as a matrix. At the end of this programme an in-built MATLAB *Fast Fourier Transform* (fft) routine is called which uses the pulse data to calculate the frequency response at the observation point. The graphical output, representing amplitude versus frequency comprises three peaks which are in complete agreement with the relative amplitudes of the input frequencies.

```
---- -------- -- -- lossless ---- --------- -- --
% lossless TLM to determine the Fourier components of an acoustic wave
% where sound speed is 300m/sec and the sampling frequency is 6kHz
% ************* INPUT PARAMETERS ******************
nmax = 20;                %the no of nodes in the problem
N=512;                    %N is the number of iterations
observation= 10;          %observation is the observation point for the fft
a=zeros(1,N);             % the observed data matrix
%***************** TLM ROUTINE *************************
%input
  for j=1:nmax
     x(j)=((pi/(0.5*nmax))*(2*j-1));
  end
   vir=0.5*sin(x)+0.1*sin(3*x)+0.05*sin(4*x);
   vil=0.5*sin(x)+0.1*sin(3*x)+0.05*sin(4*x);
for k=1:N
   %summation of incident pulses
      vtotal=vil+vir;
   %fill the observation matrix with values
      a(k)=vtotal(1,observation);
   %scatter
      vsl=vir;
      vsr=vil;
   % connect
      vil=shiftlr(vsr,1);
      vir=shiftlr(vsl,-1);
   % apply boundary conditions
      vil(1)=-vsl(1);
      vir(20)=-vsr(20);
end
ftplot
```

```
- - - - - - - - - - - - - - - - - - - - - FFT - - - - - - - - - - - - - - - - - - -
%This programme takes data from the lossless TLM routine and provides
% a frequency plot of the absolute value of the positive part of the transform
% ************************************************************
b=fft(a);
fp=b(1:N/2+1)*(1/6000);
f=6000*(0:N/2)/N;
plot(f,abs(fp))
grid
- - - - - - - - - - - - - - - - - - - - - - - - - - - - - - - - - - - - - - - - - - -
```

It is instructive to see how the number of iterations and the overall size of the mesh can affect the output from the fft. However, care should be taken to ensure that the number of iterations is a power of 2. Thus the next larger sample which the fft could accommodate without special adaptation would be 1024.

The Dispersion problem

Most numerical techniques have problems or limitations. Finite difference algorithms for parabolic equations (e.g. the diffusion equation) based on *Central Difference* schemes are always unstable. *Backward Difference* schemes are conditionally stable. Conventional lossless TLM networks are unconditionally stable, but there are some less desirable features which will now be discussed.

If the two dimensional routine shown earlier was re-written for a much larger mesh (e.g. 200 × 200) and if it was run for say 95 iterations following a single input at the centre then many anomalies would be observed. We might expect a single wave-front to propagate radially from the point of excitation. Instead we might observe the following:

- The wave-front is not radial; in some directions it seems to go slower
- The fastest propagation seems to be slower than expected
- There is much fine structure within the circle of propagation

The situation might appear even more confused if we were to have a line excitation. A single-shot input along a diagonal (figure 2.6(a) gives something that resembles a plane wave travelling in two directions. However, the same input along one of the Cartesian axes (e.g. figure 2.6(b)) gives a result that bears no similarity to a planar wave. As the co-ordinate axis and its orientation are the creations of the modeller there is obviously something wrong here.

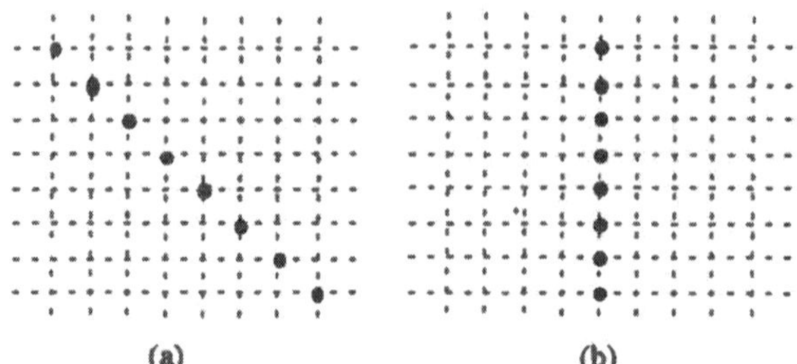

Figure 2.6 The points on a grid for single shot excitation of a planar wave: (a) diagonal excitation, (b) rectilinear excitation.

The explanation for this lies in the phenomenon of numerical dispersion, the fact that the velocity of propagation on a rectangular mesh depends on frequency and on direction. A factor 0.707 arises because of the way in which an impulse is forced to travel between neighbouring points at opposite diagonals. The distance $\sqrt{2\Delta x}$ is traversed in time $2\Delta t$, so that

$$\text{apparent velocity} = \frac{1}{\sqrt{2}}\frac{\Delta x}{\Delta t}$$

Dispersion can be presented as velocity versus frequency but within TLM it is more usual to plot velocity against a frequency equivalent, $\Delta x/\lambda$, where λ is the wavelength in question and Δx is the discretisation which is used in the model. This is shown in figure 2.7 and it can be seen that the velocity drops to zero when $\Delta x/\lambda = 0.25$ which is termed *cut-off*. This means that it is not possible to propagate a wave if the discretisation is equivalent to 4 nodes per wavelength.

It will be noted that so long as $\Delta x/\lambda \leq 0.1$ then there is not much variation in velocity and this constitutes a basis for limiting the effects of dispersion in TLM models. Any excitation/s whose frequency spectrum does not contain components with less than the equivalent of ten nodes per wavelength will not exhibit significant dispersion. So long as the modeller does not loose sight of these factors then it is possible to avoid spurious effects due to dispersion.

Figure 2.8 shows the result when a Gaussian input is used to simulate Huyghens propagation. In this case the excitation whose time dependent amplitude is given by

$$A_{\max} e^{-(t-t_0)^2/\sigma^2}$$

The value of σ depends on the required spectral range; the broader the Gaussian curve, the smaller the content of high frequency components.

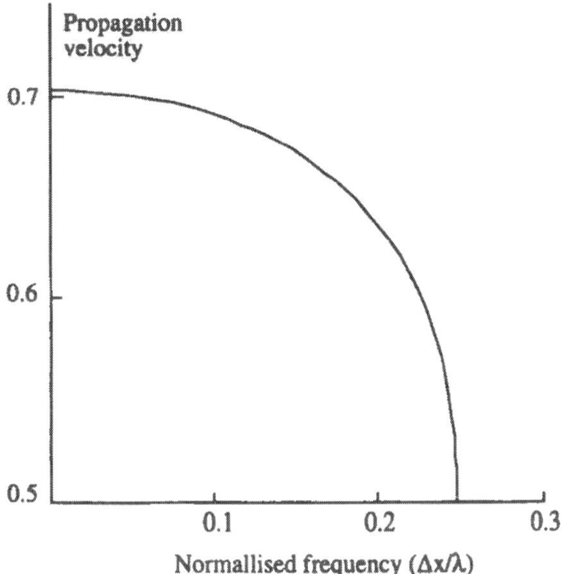

Figure 2.7 Dispersion plotted as normalised velocity versus $\Delta x/\lambda$. It will be noted that the maximum velocity is 0.707.

Thus we might decide to discretise the curve into an equivalent symmetrical histogram equivalent to 9 time-steps (figure 2.8) and the results for an input of this type are shown in figure 2.9. In this last example the excitation is placed slightly off-centre and the simulation is allowed to run until the wave-front has been reflected off the two nearest boundaries, one of which is an open-circuit and the other is a short-circuit.

Lossless TLM algorithms for materials with variable parameters

In many practical applications we may be required to treat situations where there is a change in material properties. These can generally be simulated by means of electromagnetic analogues involving variations in permittivity (ε) or permeability (μ). For instance, the propagation of sound underwater can be affected by variations in temperature or salinity level (in the case of the sea) (German U-boats during the second world war are reputed to have frequently made use of this in order to avoid detection by Allied sonar). In

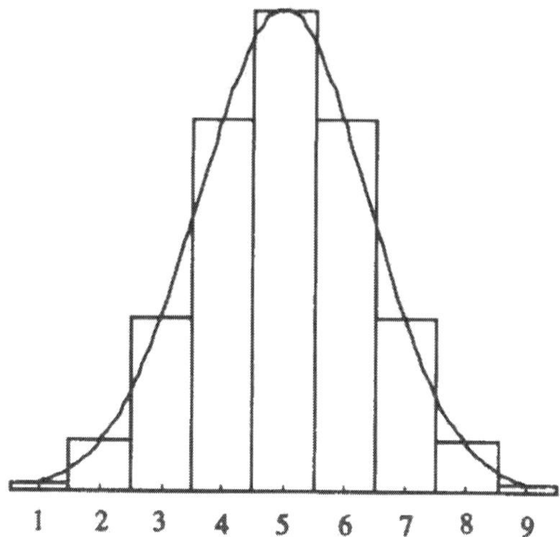

Figure 2.8 A nine-point discretised Gaussian with time-step along the horizontal axis. The vertical axis is excitation amplitude corresponding to a particular time-step.

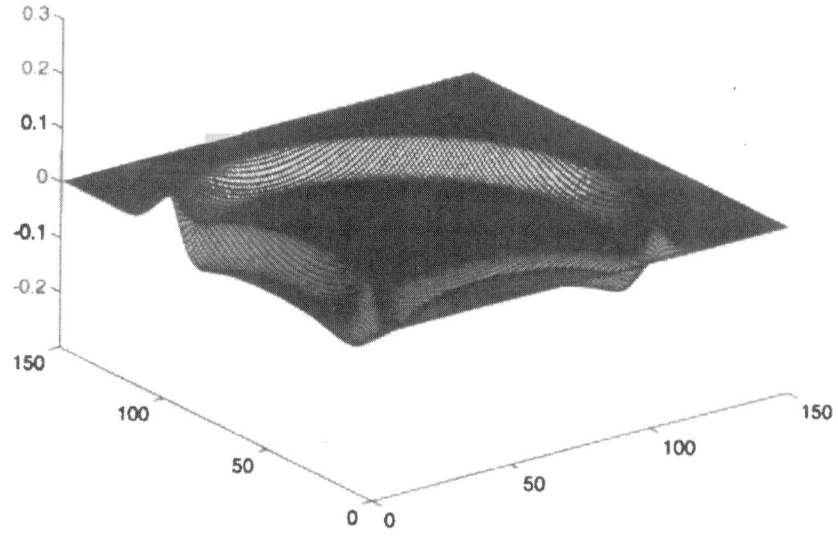

Figure 2.9 Huyghens propagation as a result of a Gaussian excitation.

the case of acoustic refraction the direct cause is the change in velocity due to change in density of the medium and it can simply be treated in TLM as analogous to a change in permittivity.

Now, if the impedance of a transmission line is given by eqn (2.13) then this could be rewritten as

$$Z = \frac{\Delta t}{\varepsilon_r C_0} \tag{2.25}$$

It can be seen that an increase in permittivity will reduce the impedance. By an identical argument we can show that in increase in permeability will lead to an increase in the impedance. Such changes give rise to additional steps in any TLM algorithm and these can be incorporated by including internodal reflections or by the use of stubs.

Internodal reflections

At the start of an iteration pulses will leave nodes (x,y) and $(x+1,y)$ as shown in figure 2.10, with magnitude being determined by eqn (2.23). They will each travel a distance $\Delta x/2$ after which they will encounter a change in impedance which will cause additional scattering. In general terms a pulse moving from its source s to an adjacent node a will experience a reflection if Z_s and Z_a are different. The reflection coefficient is given by:

$$\rho_{s\to a} \equiv \frac{Z_a - Z_s}{Z_a + Z_s} \tag{2.26}$$

The pulse arriving from a at s will experience a complementary reflection coefficient:

$$\rho_{a\to s} = \frac{Z_s - Z_a}{Z_s + Z_a} \tag{2.27}$$

The transmission coefficients are similarly defined so that the connection process across the boundary now becomes:

$$\begin{aligned} {}_{k+1}^{\;i}V_E(x,y) &= \rho_{s\to a}\,[{}_k^sE(x,y)] + \tau_{s\to a}\,[{}_k^sV_W(x+1,y)] \\ {}_{k+1}^{\;i}V_W(x+1,y) &= \rho_{a\to s}\,[{}_k^sV_W(x+1,y)] + \tau_{a\to s}\,[{}_k^sV_E(x-1,y)] \end{aligned} \tag{2.28}$$

These alterations are only relevant to pulses crossing the boundary between the two materials.

Incorporation of stub lines

There is a totally different way of approaching the problem of material discontinuities, which requires a slightly more complicated scattering

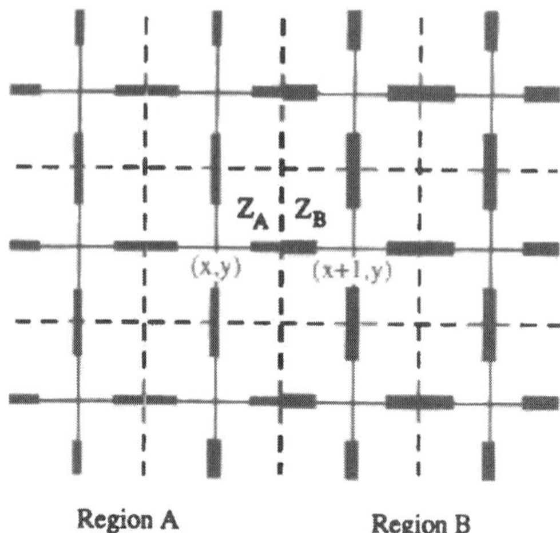

Figure 2.10 A material discontinuity half-way between two sets of nodes.

matrix but which has no requirement for an intermediate scattering of the form shown in eqn (2.28). This uses the concept of ***stubs***, which has been borrowed from microwave engineering.

If we take the two regions in figure 2.10 we can say that:

$$C_A = \varepsilon_A C_0 \quad \text{and} \quad C_B = \varepsilon_B C_0 \tag{2.29}$$

We can also say that

$$\varepsilon_B = \varepsilon_A + \varepsilon_S$$

so that

$$C_B = C_A + C_S \tag{2.30}$$

The impedance within region B is then

$$Z_B = \frac{\Delta t}{C_B} = \frac{\Delta t}{C_A + C_S} \tag{2.31}$$

This can be written in reciprocal form as:

$$\frac{1}{Z_B} = \frac{C_A}{\Delta t} + \frac{C_S}{\Delta t} = \frac{1}{Z_A} + \frac{1}{Z_S} \tag{2.32}$$

This means that Z_B can be replaced by a parallel combination of Z_A and an additional line called a ***stub*** which ensures the match between the lines as shown in figure 2.11.

The main feature of a stub is that it should act as a storage rather than a 'leakage' element. This means that data which is passed to it during the matching process is returned to the network and not lost from the network. If the stub is terminated in an open-circuit ($\rho = 1$) then the 'phase' of the data will be retained. If it is a short-circuit ($\rho = -1$) then the 'phase' of the data is inverted. A stub-line of length $\Delta x/2$ will return data to the network after one time-step: a pulse will take a time $\Delta t/2$ to reach the termination and $\Delta t/2$ to return.

In TLM algorithms it is normal to use an open-circuit terminated stub to match changes in capacitance and a short-circuit terminated stub to match changes in inductance.

The use of the half-length line introduces a small change in the definition of the stub impedance, which now becomes:

$$Z_S = \frac{\Delta t/2}{C_S} = \frac{\Delta t}{2C_S} \tag{2.33}$$

The derivation of a scattering matrix appears complicated because the circuit in figure 2.11 now has an extra component, Z_S in the load. Consequently, the scattering between connecting lines must also include scattering into *and* from the stub. This yields a matrix equation, which is in fact computationally efficient and avoids some of the complexities of the discontinuity approach.

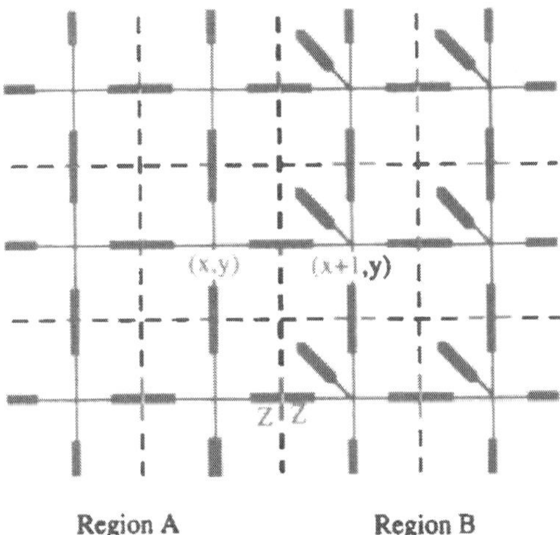

Figure 2.11 Differences in material properties loaded into stubs.

$$\begin{pmatrix} {}^sV_N \\ {}^sV_S \\ {}^sV_E \\ {}^sV_W \\ {}^sV_{st} \end{pmatrix} = \frac{1}{4Z_S + Z}\mathbb{S}\begin{pmatrix} {}^iV_N \\ {}^iV_S \\ {}^iV_E \\ {}^iV_W \\ {}^iV_{st} \end{pmatrix} \tag{2.34}$$

where

$$\mathbb{S} = \begin{pmatrix} -(Z+2Z_S) & 2Z_S & 2Z_S & 2Z_S & 2Z \\ 2Z_S & -(Z+2Z_S) & 2Z_S & 2Z_S & 2Z \\ 2Z_S & 2Z_S & -(Z+2Z_S) & 2Z_S & 2Z \\ 2Z_S & 2Z_S & 2Z_S & -(Z+2Z_S) & 2Z \\ 2Z_S & 2Z_S & 2Z_S & 2Z_S & (Z-4Z_S) \end{pmatrix}$$

In contrast to the internodal reflection method, the TLM connect process is not affected and eqn (2.24) remains applicable. However, the nodal potential is now given by

$$\phi(x) = \frac{\left[\dfrac{2\,{}^iV_N + 2\,{}^iV_S + 2\,{}^iV_E + 2\,{}^iV_W}{Z} + \dfrac{2\,{}^iV_S}{Z_S}\right]}{\left[\dfrac{4}{Z} + \dfrac{1}{Z_S}\right]} \tag{2.35}$$

The storage or delaying nature of the stub can be demonstrated by considering the one-dimensional propagation of a pulse along a lossless line which has a single half-length open-circuit stub as shown in figure 2.12. As in eqn (2.35) the equation for the nodal potential is given by the sum of the incoming currents divided by the sum of all of the impedances taken in parallel:

$$\phi = \frac{\left[\dfrac{2\,{}^iV_L + 2\,{}^iV_R}{Z} + \dfrac{2\,{}^iV_S}{Z_S}\right]}{\left[\dfrac{2}{Z} + \dfrac{1}{Z_S}\right]} \tag{2.36}$$

Initially we will consider a single pulse incident from the left (${}^iV_L = 1200$) with all others set to zero. We will also set $Z = 1$ and $Z_S = 1$ so that ${}_0\phi = 800$.

The incoming pulse sees Z_S and Z in parallel as a load and undergoes a reflection. The reflection coefficient is $-1/3$ so that -400 is reflected. Thus $(2/3)1200$ is transmitted into the other arms. Thus at the end of the scattering process we have ${}_0^sV_L = -400$, ${}_0^sV_R = 800, {}_0^sV_S = 800$.

Because there is no other scattering source present ${}_0^sV_L$ and ${}_0^sV_R$ are of no further interest. However ${}_0^sV_S$ reaches the end of the stub and is reflected back in-phase. At the end of the first time-step we then have

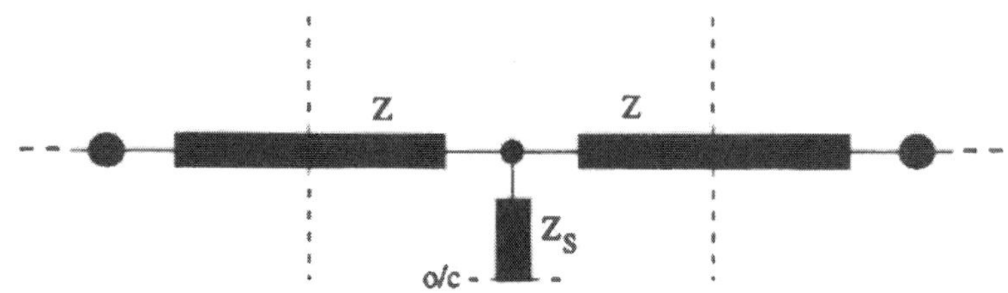

Figure 2.12 A one-dimensional lossless line with a half-length, open-circuit stub.

${}_1^iV_s = 800$ arriving back at the node which gives rise to nodal voltage ${}_1\phi = (2/3){}_1^iV_s$.

The pulse arriving at the node from the stub now sees two impedances in parallel ahead of it and therefore the reflection coefficient is $-1/3$. The pulse scattered back into the stub is then $-(1/3)800$ which becomes ${}_2^iV_s$ and thus ${}_2\phi = (2/3)[(-1/3)800]$. The next nodal voltage will then be $(2/3)$ $[(-1/3)$ $(-1/3)800]$ and so on. The process continues on for all subsequent time although as it can be seen in figure 2.13 the contributions tend to a negligible level within a few iterations.

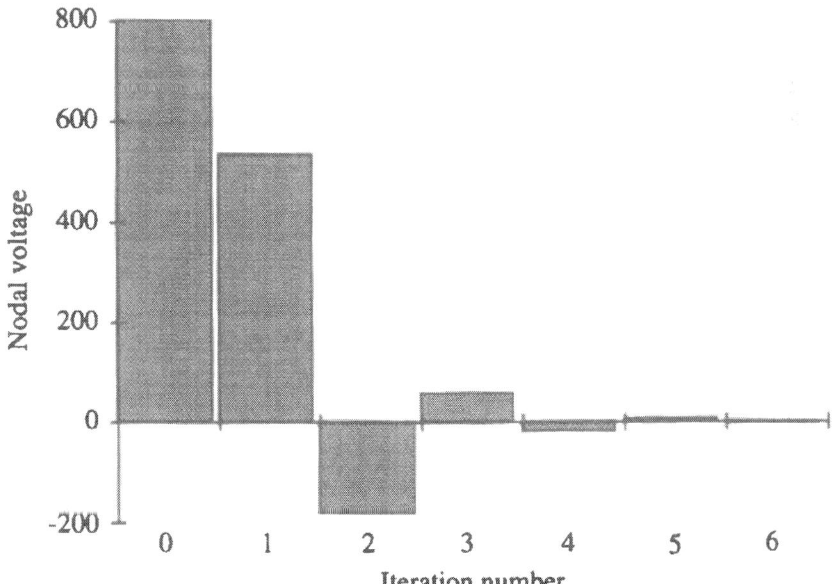

Figure 2.13 The variation of nodal voltage with time-step for an initial injection of 1200 (from the left) onto the one-dimensional mesh with a single stub ($Z_S/Z - 1$) shown in figure 2.12.

Stubs and dispersion

In another sense the action of a stub can be interpreted as leading to the effective reduction in the velocity of a wave across a mesh. As was mentioned earlier, the problem of dispersion is intimately connected with the velocity of propagation. Thus, if a stub slows down propagation it must also affect the dispersion properties. Meliani [2.8] considered a mesh in a general two-dimensional orthogonal system where $\Delta x = a$ and $\Delta y = b$. He showed that the stub capacitance could be expressed as:

$$C_s = abe\left\{1 - \frac{\Delta t^2}{\mu\varepsilon}\left[\frac{a^2+b^2}{a^2b^2}\right]\right\} \tag{2.38}$$

If we define $S = [\Delta t^2/\mu_r\varepsilon_r](a^2 + b^2\ /a^2b^2)$ with $c = 1/\sqrt{\mu_0\varepsilon_0}$ then a condition $1 - Sc^2\Delta t^2 \geq 0$ must apply to all nodes in the network, so that the unit of time discretisation is given by:

$$\Delta t = \frac{1}{c\sqrt{S}} \tag{2.39}$$

In a conventional TLM mesh $a = b$ (i.e. $\Delta x = \Delta y$), so if $\mu_r = 1$ and $\varepsilon_r = 1$ then $\Delta t = 1/c\sqrt{2}$.

If $\mu_r \neq 1$ and/or $\varepsilon_r \neq 1$ then S will be different from 2 and this analysis leads to an expression for the influence of the stub on the first cut-off:

$$\left(\frac{\Delta x}{\lambda}\right)_{\text{cut-off}} = \frac{1}{\pi}\sin^{-1}\left[\frac{1}{\Delta x\sqrt{\mu_r\varepsilon_r}\sqrt{S}}\right] \tag{2.40}$$

Thus S alters the character of figure 2.7. For example, if $S = 4$ then the maximum mesh velocity is $c/2$ and the cut-off is 0.166. This demonstrates that the bandwidth of the network is reduced as S is increased. It also indicates the need for even greater care with mesh excitation in the presence of stubs.

Conclusions

This chapter has presented the basic theory for lossless TLM in one and two dimensions. We have discussed the nature of inputs and boundaries. We have indicated the problems which can be caused by dispersion on a rectangular mesh and how they can be circumvented.

The omission of any reference to three-dimensional propagation has been intentional. The two-dimensional lossless shunt mesh which formed the basis of eqn (2.3) with $\sigma = 0$ involved magnetic fields (equivalent to currents) in the x- and y-directions with an electric field (equivalent to a voltage) in the z-direction. The theoretical basis for a three-dimensional model requires us to be able to use the other three field components E_x, E_y and H_z. These are defined by a series mesh not a shunt mesh. Several methods have been developed to link the two meshes together. Initially there was the *expanded node* [2.9] and the *asymmetric condensed node* [2.10]. The development of the *symmetrical condensed node* [2.11] means that these have now been superseded.

The one- and two-dimensional algorithms which have been developed here have immediate relevance to equivalent algorithms for thermal and particle diffusion. The applications of lossy TLM until now has been almost concerned with scalar problems (temperature, concentration etc.). The rather complicated algorithms for the condensed nodes have been developed for vector problems and are therefore not considered in this book.

References

[2.1] L.V. Bewley, "Two-dimensional fields in electrical engineering", Macmillan, New York 1948.

[2.2] G. Kron, Equivalent circuits to represent the electromagnetic field equations *Physical Review*, **64** (1943) 126–128.

[2.3] G. Kron, Equivalent circuits to the field equations of Maxwell, *Proc. IRE*, **32** (1944) 289–298.

[2.4] J. Vine, Impedance Networks chapter in "Field analysis; experimental and computation" Vitkovitch (Editor) Van Nostrand, London 1966.

[2.5] P. Hammond and J. Sykulski, "Engineering Electromagnetism; physical processes and computation" Oxford Science Publications 1994.

[2.6] C. Christopoulos, "The transmission line modelling method", OUP/IEEE Press 1995

[2.7] D. Hanselman and B. Littlefield, "Mastering MATLAB", Prentice-Hall 1996.

[2.8] H. Meliani, Mesh generation in TLM. *PhD thesis*, Nottingham University 1987, p. 97

[2.9] W.J.R. Hoefer, The transmission line matrix method – theory and applications, *IEEE Trans.* Microwave Theory and Techniques **MTT-33** (1985) 882–893.

[2.10] A. Amer, The condensed node method and its applications in transmission in power systems. *PhD thesis*, Nottingham University 1980.

[2.11] P.B. Johns, A symmetrical condensed node for the TLM method, *IEEE Trans. Microwave Theory and Techniques* **MTT-35** (1987) 370–377.

Chapter 3

The development of lossy TLM algorithms

Introduction

Within this chapter we will only consider TLM algorithms where there is a significant distributed resistance. At the most näive level we ignore the inductive term in eqn (2.7). This might seem unreasonable, but there are equivalent real-life situations where we do precisely this.

For instance, amplitude modulated (am) radio signals consist of a high-frequency carrier wave with the audio signal impressed as a modulation. The detector circuit in a radio rectifies the total signal and passes it to the amplifier. The frequency response of the amplifier is such that it recognises only the *envelope* and ignores the carrier. Even if this were not the case then the other components (the loudspeakers and our ears) would only respond to the 'slowly' changing signal.

That is precisely what is happening here; the spatial and temporal discretisations ensure that it is generally reasonable to ignore the wave-term in the Telegraphers' equation. Only in the case of ultra-fast heating will we have to do anything other than use the approximation

$$\frac{\partial^2 V}{\partial x^2} = R_d C_d \frac{\partial V}{\partial t} \tag{3.1}$$

which is equivalent to the diffusion equation with $R_d C_d = 1/D$.

As mentioned at the end of the last chapter lossless TLM is generally concerned with vector quantities. Lossy algorithms are more usually applied to scalars such as voltage and temperature. As most published work on lossy TLM up to this time has been concerned with shunt networks, the treatment which is presented here is restricted to that approach.

It may be noted that the number of references are fairly scarce in this chapter. This is because the techniques have generally been developed in connection with specific applications or with theoretical analysis. Most references are therefore held over until those chapters where they are relevant to a specific application.

Spatial discretisations and electrical networks

In lossless TLM formulations there was nothing particularly special about the spatial discretisations. The situation is somewhat different here. In the first instance we may wish to place our observation points at different locations: in the centre of the node (figure 3.1(a)) or at the interface between nodes (figure 3.1(b)). There may be very good reasons for this, such as a comparison with an experimental observation or with another numerical technique. One or other arrangement may also be more convenient if a TLM routine is to be interfaced with another technique, e.g. the effect of temperature on mechanical stress modelled using a hybrid consisting of TLM for heat-flow and finite element modelling (FEM) for the stress calculations.

The discretisations in figure 3.1 can be modelled using electrical network analogues. In a finite difference analysis of the circuit it would be possible to

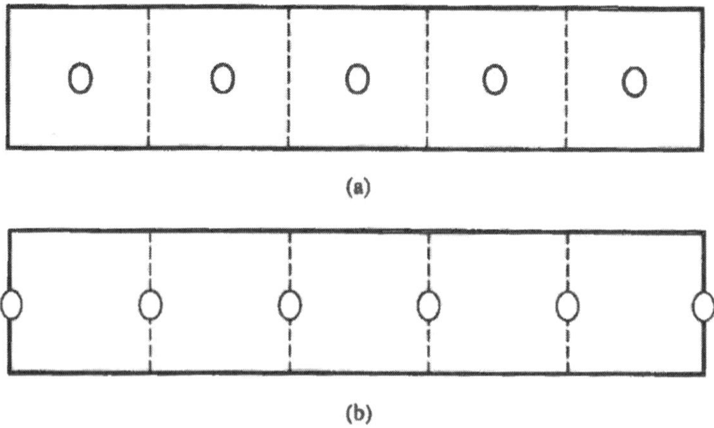

Figure 3.1 Two discretisations of a length of material showing the different positioning of observation points.

use either a T-*network* or a Π-*network* approach. The difference is shown in figure 3.2 (a), (b). Similar differences are also found in TLM implementations; and depend on the relative placement of the transmission lines and resistors within a node. If one node is separated from its neighbours by means of lengths of transmission line then this is called a *link-line* representation and is equivalent to a Π-network (figure 3.2(c)). The alternative is to have the resistors at the interface between nodes and to make observations at the centre of the transmission line. This is a *link-resistor* representation (figure 3.2(d)) and is equivalent to a T-network.

In one-dimension, the link-line and link-resistor treatments are completely equivalent, being simply the translation of the observation point and sampling interval. There are, however significant differences in two- and three-dimensional formulations which will be mentioned later.

TLM algorithm for a one-dimensional link-line nodal arrangement

There are a range of ways in which we can approach the development of a one-dimensional TLM algorithm for diffusion. In this text it is proposed to present three alternatives and to demonstrate that the outcome is equivalent for each case. The subsequent analysis will then use only one approach and it is left to the reader to adapt the others, where appropriate.

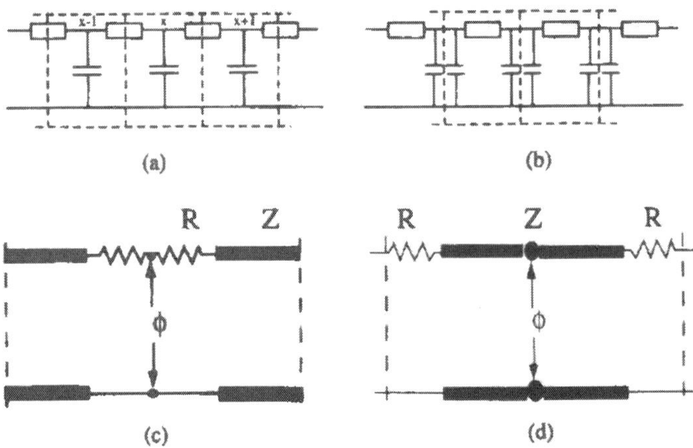

Figure 3.2 (a) T-network electrical analogue for diffusion, (b) Π-network analogue, (c) link-line TLM node, (d) link-resistor TLM node.

(1) Simple analysis

A voltage impulse entering a link-line node (figure 3.2(c)) will travel along a transmission line during a time $\Delta t/2$. At this point it encounters a discontinuity, $Z_T = (R + R + Z)$. The reflection coefficient is then:

$$\rho = \frac{R}{R+Z} \tag{3.2}$$

and the transmission coefficient is

$$\tau = \frac{Z}{R+Z} \tag{3.3}$$

Let us start by assuming that at time **k** two incident pulses, ${}^i_kV_I(x)$ and ${}^i_kV_R(x)$ are travelling along transmission lines and approaching the resistors at the centre of node **x** from **Left** and **Right** respectively (the use of boldface is only to highlight the significance of the various descriptors associated with the pulses). The Thévenin equivalent circuit assumes that these pulses have originated from voltage sources 2 ${}^i_kV_L(x)$ and 2 ${}^i_kV_L(x)$, and we can use a simple potential divider formula to calculate the contribution from each to the voltage at the centre of the node:

$${}_k\phi(x) = \frac{2\,{}^i_kV_L(x)\,(R+Z)}{(2R+2Z)} + \frac{2\,{}^i_kV_R(x)\,(R+Z)}{(2R+2Z)} = {}^i_kV_L(x) + {}^i_kV_R(x) \tag{3.4}$$

The scattering of these incident pulses (reflection and transmission) is described by:

$$\begin{aligned} {}^s_kV_L(x) &= \rho\,{}^i_kV_L(x) + \tau\,{}^i_kV_R(x) \\ {}^s_kV_R(x) &= \tau\,{}^i_kV_L(x) + \rho\,{}^i_kV_R(x) \end{aligned} \tag{3.5}$$

or

$${}^s_k\begin{bmatrix} V_L(x) \\ V_R(x) \end{bmatrix} = \begin{bmatrix} \rho & \tau \\ \tau & \rho \end{bmatrix} {}^i_k\begin{bmatrix} V_L(x) \\ V_R(x) \end{bmatrix} \tag{3.6}$$

Each of the scattered pulses now takes a time $\Delta t/2$ to travel to the boundaries of the node and a further time $\Delta t/2$ before becoming incident pulses at adjacent nodes:

$$\begin{aligned} {}^{\;\;\,i}_{k+1}V_L(x) &= {}^s_kV_R(x-1) \\ {}^{\;\;\,i}_{k+1}V_R(x) &= {}^s_kV_L(x+1) \end{aligned} \tag{3.7}$$

The repetition of eqns (3.4), (3.5) and (3.7) complete the requirements for a link-line TLM algorithm.

(2) Formal network analysis

In a strict sense the analysis outlined above has side-stepped some important issues which only become significant in two- and three-dimensional implementations or where there are additional components such as leakage (shunt) resistors located at the nodes. A formal network analysis starts by considering the summation of currents at the node and divides this by the total admittance (treating all impedances as being in parallel) to get the nodal voltage:

$$ {}_k\phi(x) = \frac{\dfrac{2\,{}_k^iV_L(x)}{(R+Z)} + \dfrac{2\,{}_k^iV_R(x)}{(R+Z)}}{\dfrac{2}{(R+Z)}} = {}_k^iV_L(x) + {}_k^iV_R(x) \tag{3.8} $$

Now the currents through the network can be expressed as:

$$ \begin{aligned} {}_kI_L(x) &= \frac{{}_k\phi(x) - 2\,{}_k^iV_L(x)}{(R+Z)} \\ {}_kI_R(x) &= \frac{{}_k\phi(x) - 2\,{}_k^iV_R(x)}{(R+Z)} \end{aligned} \tag{3.9} $$

Thus the voltage drops, $V_L(x)$ and $V_R(x)$ shown in figure 3.3 are:

$$ \begin{aligned} V_L(x) &= 2\,{}_k^iV_L(x) + {}_kI_L(x)Z = {}_k^iV_L(x) + {}_k^rV_L(x) \\ V_R(x) &= 2\,{}_k^iV_R(x) + {}_kI_R(x)Z = {}_k^iV_R(x) + {}_k^rV_R(x) \end{aligned} \tag{3.10} $$

If eqns (3.8) and (3.9) are inserted into (3.10) we obtain an identical result to eqn (3.5).

(3) Scattered current and voltage drop analysis

There is an argument which states that it is current and not voltage which is scattered in an electrical network. Eqns (2.17) and (2.18) give the reflected and transmitted currents as ${}^rI = \rho\,{}^iI$ and ${}^tI = (1-\rho){}^iI$ or $\tau\,{}^iI$. Thus the currents scattered by the link-line node in figure 3.4 are:

$$ {}_k^s\begin{bmatrix} I_L(x) \\ I_R(x) \end{bmatrix} = \begin{bmatrix} \rho & \tau \\ \tau & \rho \end{bmatrix} {}_k^i\begin{bmatrix} I_L(x) \\ I_R(x) \end{bmatrix} \tag{3.11} $$

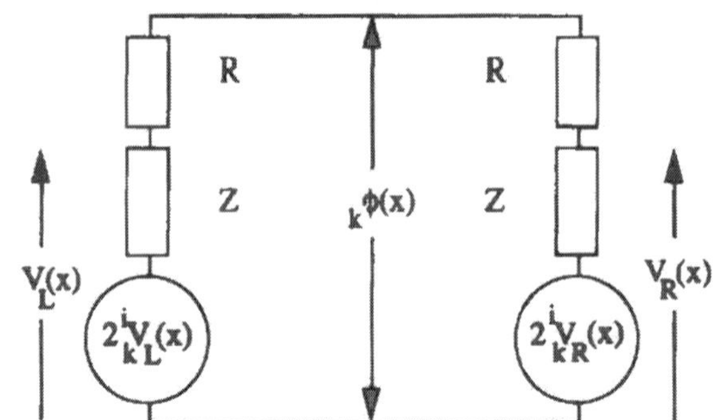

Figure 3.3 Network representation of a one-dimensional link-line node.

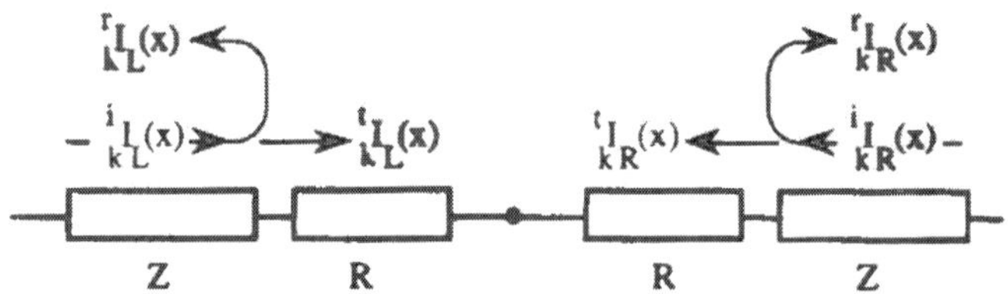

Figure 3.4 The reflection and transmission of current pulses in a link-line node.

The incident and reflected voltages on the transmission lines can be given by ${}^iIZ$ and $\rho\,{}^iIZ$. Care must be taken with the transmitted voltages which (see eqn (2.20)) are given by:

$$\begin{aligned} {}^tV_L &= (1+\rho)\,{}^iV_L = (1+\rho)Z\,{}^iI_L \\ {}^tV_R &= (1+\rho)\,{}^iV_R = (1+\rho)Z\,{}^iI_R \end{aligned} \qquad (3.12)$$

The transmitted currents give rise to voltage drops across the relevant resistors so that the super-position of voltage pulses at the centre of the node is:

$$\begin{aligned} V_L + V_R &= [{}^tV_L - R\,{}^tI_L] + [{}^tV_R - R\,{}^tI_R] \\ &= [(1+\rho)Z\,{}^iI_L - R(1-\rho)\,{}^iI_L] + [(1+\rho)Z\,{}^iI_R - R(1-\rho)\,{}^iI_R] \\ &= {}^iV_L + {}^iV_R = \phi(x) \end{aligned}$$

Once the current pulses ${}^tI_L$ and ${}^tI_L$ pass the node centre they must each traverse yet another resistor which gives rise to further voltage drops. Therefore the total scattered voltages are given by:

$$\begin{aligned} {}^sV_L &= (R+Z)[{}^sI_L - \rho\,{}^tI_L] + Z\rho\,{}^tI_L - (R+Z)[{}^sI_L - \rho\,{}^tI_L] = Z\,{}^sI_L \\ {}^sV_R &= (R+Z)[{}^sI_R - \rho\,{}^tI_R] + Z\rho\,{}^tI_R - (R+Z)[{}^sI_R - \rho\,{}^tI_R] = Z\,{}^sI_R \end{aligned}$$

Thus

$$\begin{bmatrix} {}^{s}V_L \\ {}^{s}V_R \end{bmatrix} = Z\begin{bmatrix} {}^{s}I_L \\ {}^{s}I_R \end{bmatrix} = Z\begin{bmatrix} \rho & \tau \\ \tau & \rho \end{bmatrix}\begin{bmatrix} {}^{i}I_L \\ {}^{i}I_R \end{bmatrix} = \begin{bmatrix} \rho & \tau \\ \tau & \rho \end{bmatrix}\begin{bmatrix} {}^{i}V_L \\ {}^{i}V_R \end{bmatrix} \tag{3.13}$$

which confirms that the same scattering matrix can be used for both current and voltage.

One-dimensional link-resistor formulation

The algorithm for a link resistor can also be used to give the potentials at the interface between nodes, since it is simply the summation of left- and right-going pulses. At the start of an iteration six pulses share three positions, $x-1$, x and $x+1$ which are situated at the centre of transmission lines as in figure 3.2(d) (which is the TLM equivalent of figure 3.2(a)). These are ${}_k^sV_L(x-1)$, ${}_k^sV_R(x-1)$, ${}_k^sV_L(x)$, ${}_k^sV_R(x)$, ${}_k^sV_L(x+1)$ and ${}_k^sV_R(x+1)$. The pulse which is at $x-1$ travelling to the left is no longer relevant to node x and is ignored. The same applies to the one which is travelling to the right from $x+1$. The other four pulses travel for time $\Delta t/2$ before they are scattered at the resistors. They then become incident on x from left and right as:

$$\begin{aligned} {}_{k+1}^{\;i}V_L(x) &= \rho\,{}_k^sV_L(x) + \tau\,{}_k^sV_R(x-1) \\ {}_{k+1}^{\;i}V_R(x) &= \rho\,{}_k^sV_R(x) + \tau\,{}_k^sV_L(x+1) \end{aligned} \tag{3.14}$$

The pulses arrive simultaneously at x and they sum to give the instantaneous potential

$${}_{k+1}\phi(x) = {}_{k+1}^{\;i}V_L(x) + {}_{k+1}^{\;i}V_R(x) \tag{3.15}$$

Once the pulses pass on their way after this incidence it is useful to re-designate them for use at the next iteration

$$\begin{aligned} {}_{k+1}^{\;s}V_R(x) &= {}_{k+1}^{\;i}V_L(x) \\ {}_{k+1}^{\;s}V_L(x) &= {}_{k+1}^{\;i}V_R(x) \end{aligned} \tag{3.16}$$

A complete algorithm consists of the repetition of eqns (3.14) – (3.16) for k iterations, where $k\Delta t$ is the total time of the simulation.

Boundaries

As in lossless TLM, the standard descriptions of short-circuit and open-circuit terminations can be used to model certain classes of physical boundaries in heat and matter simulations.

(1) An insulating boundary

Any heat approaching an insulating boundary is reflected back into the physical problem. This is the open-circuit ($\rho = 1$) condition and in both link-line and link resistor formulations it is customary to place it at the interface between nodes. Thus, a pulse travelling from a node during time $\Delta t/2$, encounters the boundary and arrives back at the node at the end of the time-step.

(2) A symmetry boundary

As in lossless propagation, the computation of heat or matter profiles can sometimes be reduced by exploiting any symmetry in the problem, so that only half the problem need be simulated. The interface along the symmetry axis becomes an open-circuit boundary ($\rho = 1$).

(3) A perfect heat-sink boundary

This is covered by the definition of the short-circuit boundary of lossless TLM, but some additional care is required. Once again, the boundary is placed at the interface between two nodes. However, there are slight differences between the link-line and link-resistor formulations. In a link-line

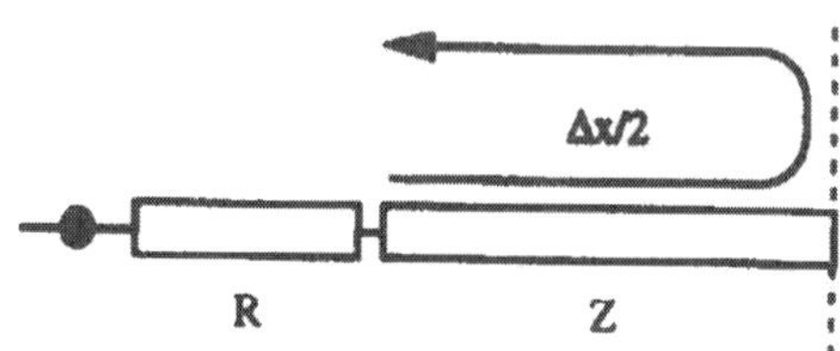

Figure 3.5 Reflection at an insulating boundary.

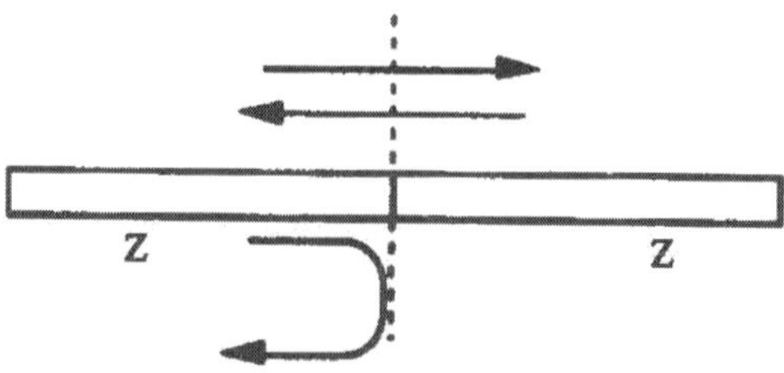

Figure 3.6 The equivalence between a symmetry boundary and an insulating boundary. Two identical pulses crossing this boundary are identical to a reflected pulse.

model the pulse is half-way along a transmission line when it sees a termination $Z_T = 0$ in front of it. The reflection coefficient is thus $\rho = -1$. In a link resistor node the normal load impedance which a pulse sees as it reaches the end-of-the-line is $R + R + Z$. One of these two resistors is associated with the node. The normal description of a short circuit condition in such cases is that the short is located immediately outside the node. Thus the line terminating impedance is $Z_T = R$ and the reflection coefficient from a short-circuit is given by:

$$\rho = \frac{R - Z}{R + Z} \tag{3.17}$$

(4) Constant temperature boundaries

In a link-line model the transmission line touches the boundary, whose value is held constant (V_C, as shown in figure 3.7).

We will assume that there is a 'ghost' node outside the boundary, and that this has a source and transmission line. This ensures that the value of the potential (V_C) at the surface, which is the summation of the pulse incident on node 1 at each new time-step and the pulse scattered from node 1 at the previous time-step is always constant.

$$_{k+1}^{\ i}V_L(1) + {}_k^sV_L(1) = V_C \tag{3.18}$$

Since ${}_k^sV_L(1)$ is known at the present time-step, this equation can be used to calculate $_{k+1}^{\ i}V_L(1)$.

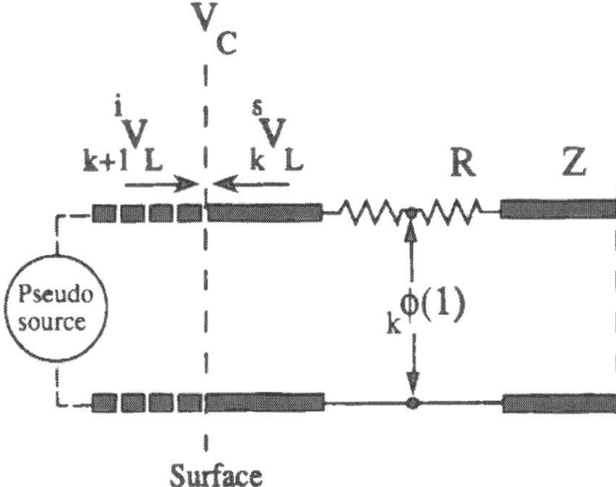

Figure 3.7 The boundary node in a link-line formulation showing the 'ghost' source and line together with the incident and scattered pulses.

The situation with a link-resistor treatment is quite different since it is the resistor, not the transmission line which touches the boundary. There are then two separate considerations:

- the input from the source which can now be situated *at* the boundary (figure 3.8(a)),
- the history of the pulse which is scattered from node 1 and which *approaches* the boundary (figure 3.8(b)).

The source (V_C) on the boundary sees a series connection of resistor and impedance, so that the standard potential divider formula gives the signal injected into the line. The pulse scattered towards the boundary sees a short-circuit so that the total which is incident from the left at a new time-step is the sum of these contributions.

$$ {}_{k+1}^{\ i}V_L(1) = V_C \frac{Z}{R+Z} + {}_k^s V_L(1) \frac{R-Z}{R+Z} \tag{3.19} $$

Inputs

We now turn to various methods of excitation which closely model physical processes. The initial treatments are for excitation (pulse injection) into boundary-free (bulk) material. The use of impulsive inputs in lossy TLM does not normally present the same problems as it does in lossless TLM. The dispersion due to the presence of resistors is such that it swamps out any other effect (except perhaps during the first few time-steps [3.1]).

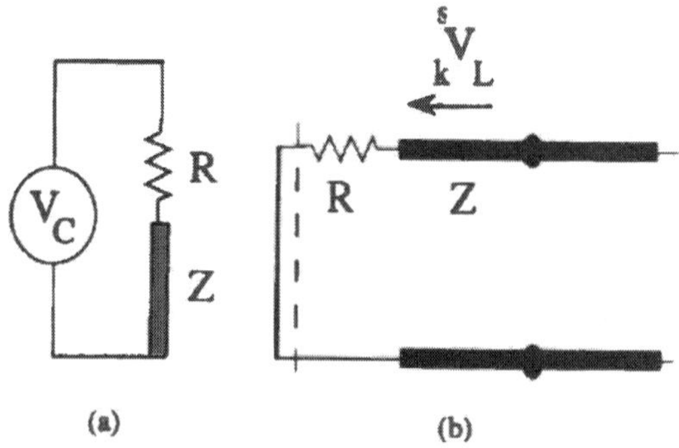

Figure 3.8 (a) the network which is seen by the input at the boundary (b) the situation which is seen by the pulse scattered towards the boundary.

Single shot injection into bulk material

This description covers inputs which ultimately lead to Gaussian distributions of matter, heat or temperature profile. It consists of a voltage source (equivalent to temperature input) or a current source (equivalent to heat input) which is switched across one node point during the first iteration in the simulation. If we were to take the viewpoint of the signal coming from the excitation source then it is clear from figure 4.22 that it sees a junction with equal impedance to left and to right.

Thus the product, $I_{EX}\Delta t$ represents a charge which models the single-shot injection of impurities into a medium. The current divides into equal quantities moving to left and to right. The subsequent diffusion of these current pulses could be monitored using TLM. However the introductory treatment that has been given in this chapter has been expressed entirely in terms of voltage. This is not a problem, since a current $I_{EX}/2$ passing through $R + Z$ on the left will give rise to a voltage ${}^{i}V_L$. Similarly there will be a voltage ${}^{i}V_R$ moving to the right.

In terms of simple diffusion we could say that if 1000 particles were injected at the source then the initial conditions would be: ${}_{k=0}^{\;\;i}V_L = 1000/2$ and ${}_{k=0}^{\;\;i}V_R = 1000/2$. This would be sufficient to initiate a diffusion algorithm.

At this point it is worth considering a curiosity of the link-line algorithm when the propagation of a single shot input is being simulated.

At time $k = 0$ an excitation at x gives rise to pulses travelling left and right. They take time Δt to reach adjacent nodes, so that at $k = 1$ the values of $\phi(x-1)$ and $\phi(x+1)$ are defined but $\phi(x)$ is not. Similarly, the pulses scattered at $(x+1)$ and $(x-1)$ scatter so that at $k = 2$ the values $\phi(x)$, $\phi(x-2)$ and $\phi(x+2)$ are defined while $\phi(x+1)$ and $\phi(x-1)$ are not. This will manifest itself as apparent jumps to zero of any node at alternate time steps, which is obviously unphysical. Such behaviour is

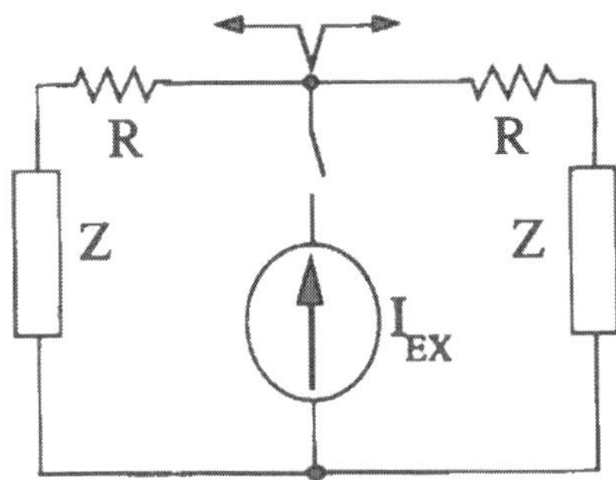

Figure 3.9 The equal distribution of the output from an excitation source

observed in other (non-TLM) techniques and in fact represents a singularity. This can be observed in the simple piece of code to model the diffusion in a one-dimensional link-line network during the first few iterations following a single input into the centre of the space.

```
*****************************************************************
#include <stdio.h>
#define refl 0.5
#define trans 0.5
void main(void)
{
    float voltage [22]={0},
          vi_left [22]={0},
          vi_right [22]={0},
          vs_left [22]={0},
          vs_right [22]={0};
    int i, j, k;
    /* print heading */
    printf("\n\nLink line:Results over five iterations for nodes 7 to 15");
    printf("/n===================================\n");
    /* input */
    vi_left[11]=256;
    vi_right[11]=256;
    /* iteration */
    for(i=1; i <= 5; i++)
    {
        /* summation */
        for(j=1; j <= 21; j++) voltage[j]=vi_left[j] + vi_right[j];
        /* print results */
        printf("\n");
        for(j=7; j<= 15; j++) printf("% 8.3f",voltage[j]);
        /* scattering */
        for(j=1; j<= 21; j++)
        {
          vs_left[j]=refl * vi_left[j] +trans * vi_right[j];
          vs_right[j]=refl * vi_right[j] + trans * vi_left[j];
        }
        /* connect */
        for(j=2; j<=21; j++) vi_left[j]=vs_right[j-1];
        for(j=1; j<= 20; j++) vi_right[j] =vs_left[j+1];
    }
}
*****************************************************************
```

This phenomenon is not observed in a link resistor algorithm. The point is clearly demonstrated by running the equivalent programme for the same one-dimensional problem.

```
************************************************************************
#include <stdio.h>
#define refl 0.5
#define trans 0.5
void main(void)
{
    float voltage[22]={0},
      vi_left[22]={0},
      vi_right[22]={0},
      vs_left[22]={0},
      vs_right[22]={0};
    int i, j, k;
    /* print heading */
    printf("\n\n Link resistor:Results over five iterations for nodes 7
         to 15");
    printf("\n==================================\n");
    /* input */
    vi_left[11]=256;
    vi_right[11]=256;
    /* iteration */
    for(i=1; i <= 5; i++)
    {
      /* summation */
      for(j=1; j<=21; j++) voltage[j]=vi_left[j] + vi_right[j];
      /* print results */
      printf("\n");
      for(j=7; j<=15; j++) printf("%8.3f",voltage[j]);
      /* scattering */
      for(j=1; j<=21; j++) vs_left[j]=vi_right[j];
      for(j=1; j<=21; j++) vs_right[j]=vi_left[j];
      /* connect */
      for(j=2; j<= 20; j++)
      {
        vi_left[j]=refl * vs_left[j]+trans * vs_right[j-1];
        vi_right[j]=refl * vs_right[j] + trans * vs_left[j+1];
      }
    }
}
************************************************************************
```

In general we can say that 'jumps-to-zero do not happen:

- in link-resistor formulations.
- if some of the capacitance is concentrated at each node centre as a stub
- if the excitation point is moved to the boundary between two nodes
- in most case of multiple excitations at a point.

Some aspects of this will be revisited in the next chapter when we come to discuss the theory of lossy TLM.

Multiple and constant injection into bulk material

(1) Constant heat source

A constant rate of heating means that an additional input will be required at every time-step. It is possible to arrange the input at any point in a TLM formulation, but it is often most convenient to add a pulse (I_{EX}) immediately after the incident step at each iteration so that the nodal potential is:

$$ {}_{k+1}\phi(x) = \frac{\left[\frac{2\,{}_{k+1}{}^{i}V_L(x)}{(R+Z)}\right] + \left[\frac{2\,{}_{k+1}{}^{i}V_R(x)}{(R+Z)}\right] + I_{EX}}{\left[\frac{2}{(R+Z)}\right]} \tag{3.20} $$

The scattered pulses are then:

$$ \begin{aligned} {}^{s}V_L(x) &= \rho\,{}^{i}V_L(x) + \tau\,{}^{i}V_R(x, y+1) + \frac{V_{EX}}{2} \\ {}^{s}V_R(x) &= \tau\,{}^{i}V_L(x) + \rho\,{}^{i}V_R(x, y+1) + \frac{V_{EX}}{2} \end{aligned} \tag{3.21} $$

where $V_{EX} = I_{EX}\,(R + Z)$

In a situation where an input is time varying this will simply involve the adjustment of the amplitude of excitation at every time-step.

Figure 3.10 shows an interesting curiosity which is due to an unphysical definition of input and boundary. Let us suppose that we have a single injection of 1000 at $k=0$. The 500 which moves to the left immediately encounters a reflecting boundary and is returned just as the other 500 arrives at node 2. This 'diffusion by pairs' repeats itself thereafter and can be considered as the 'fold-back of a Gaussian curve which is displaced by a distance Δx.

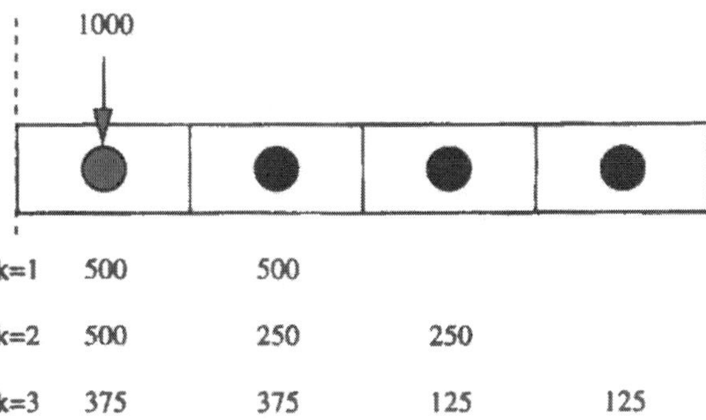

Figure 3.10 'Diffusion by pairs'. The values at the first four nodes during the first three iterations following a single-shot input of 1000 at node 1 are shown.

The extension to two and three dimensions link-line formulations

Figure 3.11 shows a node with a pulse incident from the west at the instant before it is reflected. The impedance which the pulse sees at the discontinuity is a resistor in series with a parallel arrangement of three impedances $(R + Z)$. The reflection coefficient is therefore

$$\rho = \frac{3R + R + Z - 3Z}{3R + R + Z + 3Z} = \frac{2R - Z}{2R + 2Z} \tag{3.22}$$

The transmitted component down any arm is τ, where $3\tau = 1 - \rho$.

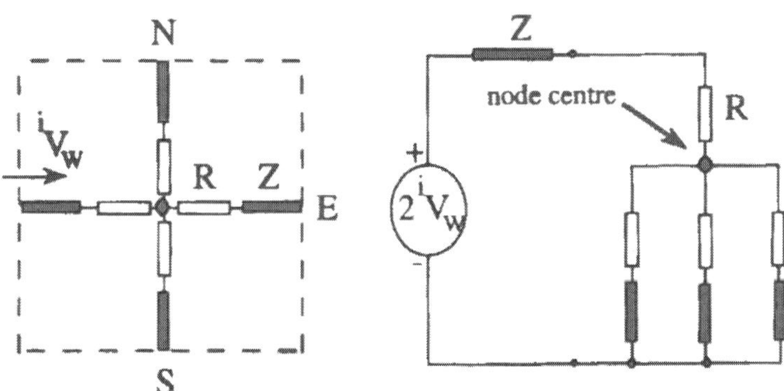

Figure 3.11 A two-dimensional link-line node and its lumped equivalent circuit with a pulse incident from the west.

The lumped circuit due to a single incident pulse is also shown in figure 3.11, but we can assume that in a general formulation there are pulses incident from all directions. Therefore the potential at (x, y) is best expressed as the current from each contributor passing through the point divided by the total admittance (reciprocal of impedance) of the node:

$$
\begin{aligned}
{}_k\phi(x,y) &= \frac{\left[\dfrac{2\,{}_k^iV_N(x,y)}{R+Z}+\dfrac{2\,{}_k^iV_S(x,y)}{R+Z}+\dfrac{2\,{}_k^iV_E(x,y)}{R+Z}+\dfrac{2\,{}_k^iV_W(x,y)}{R+Z}\right]}{\left[\dfrac{4}{R+Z}\right]} \\
&= \frac{{}_k^iV_N(x,y)+{}_k^iV_S(x,y)+{}_k^iV_E(x,y)+{}_k^iV_W(x,y)}{2}
\end{aligned}
\tag{3.23}
$$

The pulse which is scattered from each arm consists of one reflected and three transmitted components.

$$
{}_k\begin{pmatrix} {}^sV_N \\ {}^sV_S \\ {}^sV_E \\ {}^sV_W \end{pmatrix} = \begin{pmatrix} \rho & \tau & \tau & \tau \\ \tau & \rho & \tau & \tau \\ \tau & \tau & \rho & \tau \\ \tau & \tau & \tau & \rho \end{pmatrix} {}_k\begin{pmatrix} {}^iV_N \\ {}^iV_S \\ {}^iV_E \\ {}^iV_W \end{pmatrix} \tag{3.24}
$$

The connection process is exactly as in lossless TLM.

$$
\begin{aligned}
{}_{k+1}^{\;\;i}V_N(x,y) &= {}_k^sV_S(x,y+1) \\
{}_{k+1}^{\;\;i}V_S(x,y) &= {}_k^sV_N(x,y-1) \\
{}_{k+1}^{\;\;i}V_E(x,y) &= {}_k^sV_W(x+1,y) \\
{}_{k+1}^{\;\;i}V_W(x,y) &= {}_k^sV_E(x-1,y)
\end{aligned}
\tag{3.25}
$$

Up to now we have used the points of a compass to designate the direction of scattering *to* and incidence *from*. This starts to get out of hand in three-dimensional formulations although we could use N,S,E,W with U for up and D for down. Electromagnetic modellers tend to use direction numbers and this is certainly the only way when you have got to keep track of 18 or more pulses. In this presentation we only have 6 scattered and 6 incident pulses, but nevertheless, it is beneficial to get used to direction numbers.

We can easily develop the nodal voltage from the equivalent in two dimensions:

$$
{}_k\phi(x,y,z) = \frac{1}{3}\sum_{j=1}^{6} {}_k^iV_j(x,y,z) \tag{3.26}
$$

The scattering process can also be extended from two dimensions. Because the node is a centre of symmetry it is possible to write:

$$
{}_k\begin{pmatrix} {}^sV_1 \\ {}^sV_2 \\ {}^sV_3 \\ {}^sV_4 \\ {}^sV_5 \\ {}^sV_6 \end{pmatrix} = \begin{pmatrix} \rho & \tau & \tau & \tau & \tau & \tau \\ \tau & \rho & \tau & \tau & \tau & \tau \\ \tau & \tau & \rho & \tau & \tau & \tau \\ \tau & \tau & \tau & \tau & \rho & \tau \\ \tau & \tau & \tau & \tau & \tau & \rho \end{pmatrix} {}_k\begin{pmatrix} {}^iV_1 \\ {}^iV_2 \\ {}^iV_3 \\ {}^iV_4 \\ {}^iV_5 \\ {}^iV_6 \end{pmatrix} \tag{3.24}
$$

The reflection coefficient is similarly derived from eqn (3.22) as:

$$
\rho = \frac{3R - 2Z}{3R + 3Z} \text{ with } 5\tau = 1 - \rho \tag{3.28}
$$

It is only when we come to the connection process that there is a need to be up-front about the direction indices. One possible choice is shown in figure 3.12 and it should be easy to deduce the set of connection equations from this.

Link-resistor formulations

In a two-dimensional link-resistor node the transmission lines intersect as in figure 3.13. This gives rise to scattering at discrete time intervals which are

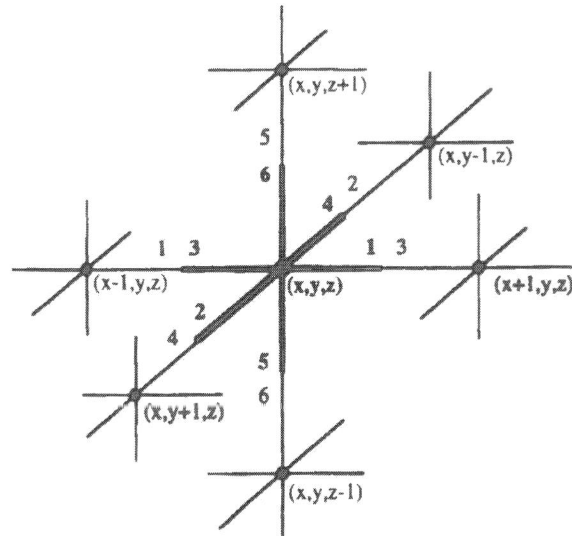

Figure 3.12 An arbitrary choice of direction indices (shown in bold) with the connection to neighbouring nodes in three-dimensions.

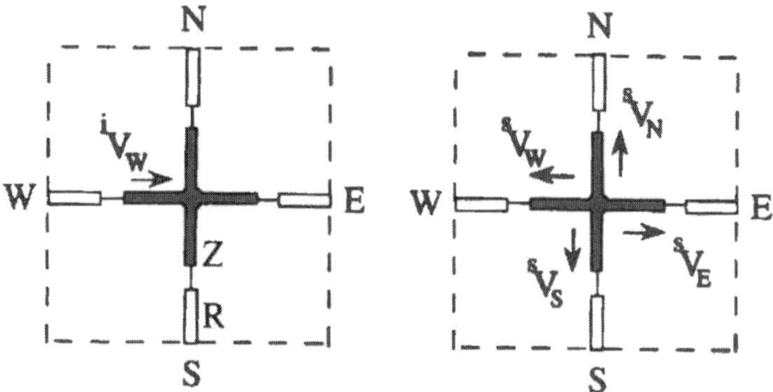

Figure 3.13 The two scattering events in a link-resistor cell separated in space by $\Delta x/2$ and in time by $\Delta t/2$.

identical to those lossless TLM namely:

$$\begin{pmatrix} {}^sV_N \\ {}^sV_S \\ {}^sV_E \\ {}^sV_W \end{pmatrix}_k = \frac{1}{2}\begin{pmatrix} -1 & 1 & 1 & 1 \\ 1 & -1 & 1 & 1 \\ 1 & 1 & -1 & 1 \\ 1 & 1 & 1 & -1 \end{pmatrix} {}_k\begin{pmatrix} {}^iV_N \\ {}^iV_S \\ {}^iV_E \\ {}^iV_W \end{pmatrix} \tag{3.29}$$

However the presence of linking resistors gives rise to a second set of four scattering events which occur at the half time intervals. These are incorporated into the connection equations:

$$\begin{aligned} {}_{k+\frac{1}{2}}^{\;i}V_N(x,y) &= \rho' {}_k^sV_N(x,y) + \tau' {}_k^sV_S(x,y+1) \\ {}_{k+\frac{1}{2}}^{\;i}V_S(x,y) &= \rho' {}_k^sV_S(x,y) + \tau' {}_k^sV_N(x,y-1) \\ {}_{k+\frac{1}{2}}^{\;i}V_E(x,y) &= \rho' {}_k^sV_E(x,y) + \tau' {}_k^sV_W(x+1,y) \\ {}_{k+\frac{1}{2}}^{\;i}V_W(x,y) &= \rho' {}_k^sV_W(x,y) + \tau' {}_k^sV_W(x-1,y) \end{aligned} \tag{3.30}$$

where $\rho' = R/(R+Z)$ and $\tau' = Z/(R+Z)$. This is because the pulse reflected from the node and travelling along a line sees its resistor and the resistor and transmission line of the target node as a mis-matching load.

The nodal potential is then

$$ {}_{k+1}\phi(x,y) = \frac{1}{2}\sum_{j=1}^{4} {}_{k+\frac{1}{2}}^{\;i}V_j(x,y) \tag{3.31}$$

The extension of the link-resistor formulation to three-dimensions can generally be done as it was in the link-line case. The scattering equations are:

$$
{}_k\begin{pmatrix} {}^sV_1 \\ {}^sV_2 \\ {}^sV_3 \\ {}^sV_4 \\ {}^sV_5 \\ {}^sV_6 \end{pmatrix} = \begin{pmatrix} \rho & \tau & \tau & \tau & \tau & \tau \\ \tau & \rho & \tau & \tau & \tau & \tau \\ \tau & \tau & \rho & \tau & \tau & \tau \\ \tau & \tau & \tau & \rho & \tau & \tau \\ \tau & \tau & \tau & \tau & \rho & \tau \\ \tau & \tau & \tau & \tau & \tau & \rho \end{pmatrix} {}_k\begin{pmatrix} {}^iV_1 \\ {}^iV_2 \\ {}^iV_3 \\ {}^iV_4 \\ {}^iV_5 \\ {}^iV_6 \end{pmatrix} \tag{3.32}
$$

Nodal scattering occurs when an impulse arrives along a transmission line and sees 5 transmission lines in parallel at the node centre. Thus, by extension from two-dimensions we can show that $\rho = -2/3$ and $\tau = 1/3$.

The connection equations can be derived from figure 3.12 but they will include reflections and transmissions with coefficients $\rho' = R/(R+Z)$ and $\tau' = Z/(R+Z)$. Taking just one as an example:

$$
{}_{k+1}^{\ \ i}V_1(x,y,z) = \rho' \, {}_k^sV_1(x,y,z) + \tau' \, {}_k^sV_3(x+1,y,z) \tag{3.33}
$$

In conclusion, it can be seen that as the dimensionality goes up there is significantly more computation in a link-resistor node and for this reason link-line formulations are the more favoured approach.

Non-uniformities in mesh and material properties

In the treatment of a real problem we may have spatial variations in the value of Z and/or R. These parameters may be functions of position, time or the magnitude of the nodal potential. Variations as a function of position may reflect differences in adjacent materials i.e. material inhomogeneities. These can also be due to the manner in which we choose to mesh a particular problem: a cylindrical geometry is best treated by means of a polar mesh, but this does mean that the volume and area of adjacent nodes is quite different. Finally, there are many instances in thermal problems where the value of temperature at any position can have a strong influence on the local specific heat and thermal conductivity. The remainder of this chapter will be devoted to techniques for handling non-uniformities, although the historical development and application to real problems will be considered in chapter 5.

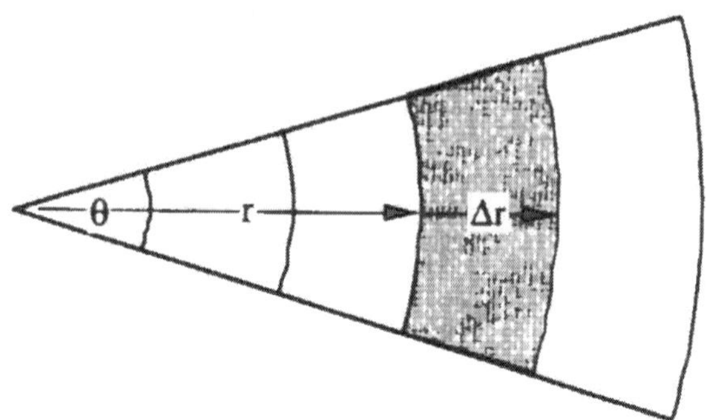

Figure 3.14 Nodes of equal length in a polar mesh.

Stubs or internodal reflections

Polar meshes for lossless TLM models with axial symmetry were first developed by Al-Mukhtar and Sitch [3.2] and examined in detail by Naylor [3.3]. de Cogan and John [3.4] extended the technique to diffusion problems. In this section we will examine an aspect of the problem before looking at non-uniform meshes in greater detail later.

Each node in a cylindrical problem has capacitance given by the product of material density, volume and specific heat. The thermal resistance given by the length divided by the product of area and thermal conductivity. The volume may be calculated as the nodal area (defined by integration) times the nodal height (Δh). There is however a question when we come to calculate the area for radial flux. $\Delta h(\theta(r+\Delta r))$ gives an over-estimate, while $\Delta h(\theta r)$ is an under-estimate. Most TLM modellers use the mean value.

The outer boundary (the right side in figure 3.14) can be defined in an arbitrary way, but there is a problem with the treatment of the extremity on the left. It is conventional to treat this as a ($\rho = 1$) boundary and this can be explained in two ways.

(i) The area corresponding to the radial centre is obviously zero and therefore the resistance of the point must be infinite. $\rho = 1$ follows from an electromagnetic definition of reflection coefficient.

(ii) The radial centre in a homogeneous problem is also a centre of symmetry. Thus whatever flows from right to left through the point must be matched by an equal flow from left to right.

To some extent these are intuitive arguments and Cacoveanu *et al.* [3.5] have adopted a more rigorous approach which places a finite sized node about the centre. This has $2n$ branches (to ensure continuity of the radial lines) and an open-circuit stub (to ensure pulse synchronism).

If the segment of a polar mesh as shown in figure 3.14 were part of a model for thermal conduction with unit depth then the thermal capacitance and hence the impedance ($\Delta t/C$) would be greater as the nodes moved outward from the centre. As it stands there would be an impedance mismatch between nodes in a link-line formulation. This would not be the case if a link-resistor scheme were used. Alternatively we could distribute the capacitance unevenly around the node so that linking transmission lines in adjacent nodes would be matched. Of course this would be equivalent to moving the observation point from the centre of the node.

Instead of moving the observation point we could choose to have an uneven mesh as shown in figure 3.15. In this case the volume of each node is now identical so that there are no link-line mismatches [3.6].

The uniform mesh of figure 3.14 can always be used with stubs. The innermost node would represent the smallest capacitance in a thermal conduction problem. It is therefore used as the base-line value, C_B: the capacitances of the successive nodes are then $C_B + \Delta C_1$, $C_B + \Delta C_2$, $C_B + \Delta C_3$ etc. The component C_B is used as the link-line in each node and the other part is incorporated as a half-length, open-circuit stub at the centre of each node. The value of the stub is given by:

$$Z_S = \frac{\Delta t}{2\Delta C} \tag{3.34}$$

The derivation of the TLM algorithm for the stub-loaded mesh follows the same lines as eqns (2.33) – (2.35) but the presence of the resistors increases the complexity of the derivation. In a one-dimensional formulation we now have pulses from **L**eft, **R**ight and **S**tub.

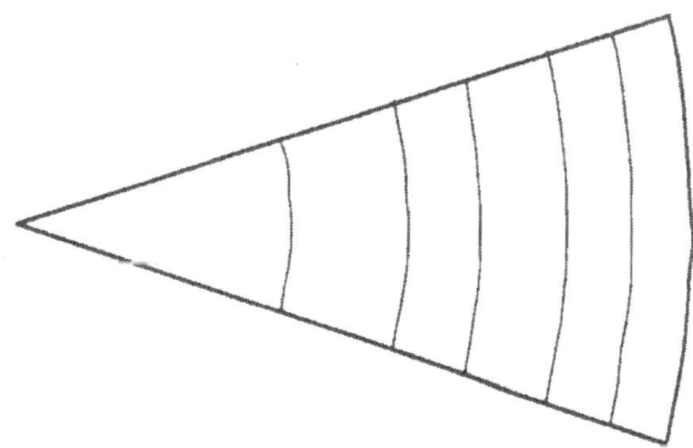

Figure 3.15 A set of nodes of equal surface area in a polar mesh.

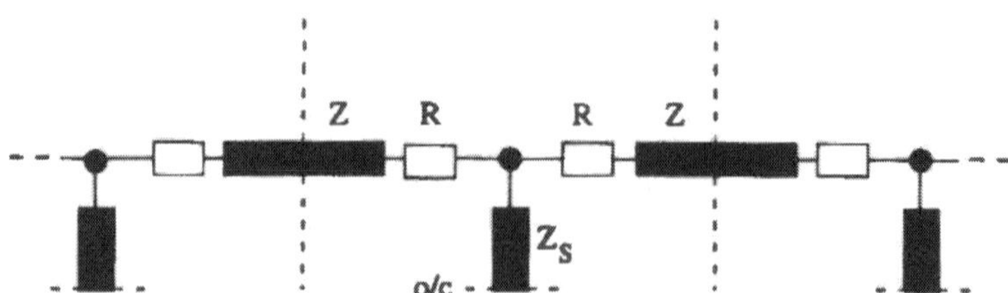

Figure 3.16 A one-dimensional mesh with stubs.

Inspection of figure 3.16 shows that pulses from left and right will be reflected with a coefficient:

$$\rho = \frac{2RZ_s + (R+Z)(R-Z)}{(R+Z)(R+Z+2Z_s)} = 1 - \tau \tag{3.35}$$

τ is the component of the incoming pulse that is transmitted into the arm on the other side of the node. The incoming component that transfers to the stub is given by:

$$\tau' = \frac{R+Z}{R+Z+Z_s}\tau = \frac{2Z}{R+Z+2Z_s} \tag{3.36}$$

A pulse which travels along the stub towards the node will see two identical arms at the discontinuity and will have a reflection coefficient:

$$\rho_s = \frac{Z+R-2Z_s}{Z+R+2Z_s} = 1 - \tau_s. \tag{3.37}$$

The scattering process are then given in matrix form as:

$${}_k\begin{pmatrix} {}^sV_L \\ {}^sV_R \\ {}^sV_S \end{pmatrix} = \begin{pmatrix} \rho & \tau & \tau_s \\ \tau & \rho & \tau_s \\ \tau' & \tau' & \rho_s \end{pmatrix} {}_k\begin{pmatrix} {}^iV_L \\ {}^iV_R \\ {}^iV_S \end{pmatrix} \tag{3.38}$$

The normal connection equations are used with the addition of the incidence due to the open-circuit termination in the stub: ${}_{k+1}^{\;\;i}V_S(x) = {}_k^sV_S(x)$

The nodal potential is then

$${}_{k+1}\phi(x) = \frac{\dfrac{2\,{}_{k+1}^{\;\;i}V_L(x)}{(R+Z)} + \dfrac{2\,{}_{k+1}^{\;\;i}V_R(x)}{(R+Z)} + \dfrac{2\,{}_{k+1}^{\;\;i}V_S(x)}{Z_S}}{\dfrac{2}{(R+Z)} + \dfrac{1}{Z_S}} \tag{3.39}$$

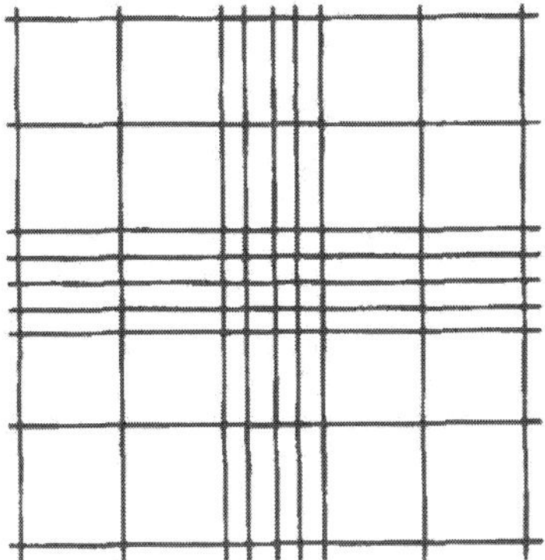

Figure 3.17 Saguet/Pic mesh refinement.

Variable mesh algorithms

In the previous section there was mention of the use of an irregular mesh to overcome a particular problem [3.7]. There are frequently other reasons why irregular meshes should be used or why it may be necessary to connect different meshes together. Several TLM researchers have developed algorithms which are detailed below.

In 1981 Saguet and Pic [3.8] devised a scheme for the refinement of a rectangular mesh in the form shown in figure 3.17. This naturally raised the problem of pulse synchronism. de Cogan *et al* [3.6] drew inspiration from this and developed a similar technique which was applied to one-dimensional heat-flow problems. They considered a uniform material where one node was longer than another by a factor, b, that is it had b times the resistance and b times the capacitance (figure 3.18).

The impedances are given by:

$$Z_A = \frac{\Delta t}{\Delta x C_d} \quad Z_B = \frac{\Delta t}{b \Delta x C_d}$$

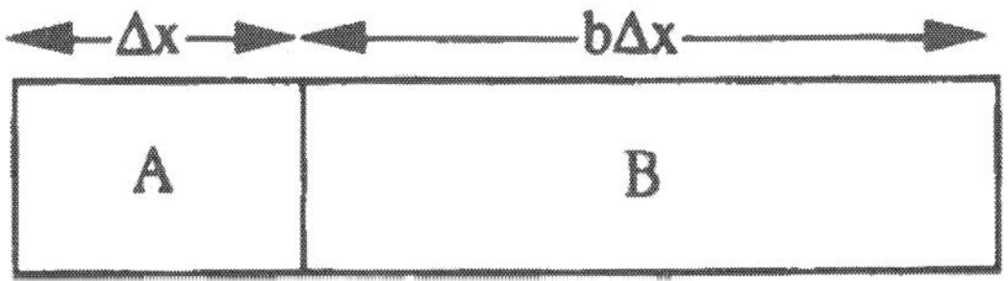

Figure 3.18 The connection of nodes of unequal length.

Thus

$$Z_B = \frac{Z_A}{b}$$

It may be of interest that de Cogan and Enders [3.9] subsequently adopted a variant of this process to provide impedance matching across adjacent materials with different thermal properties.

These developments have not been limited to one-dimension. The Pulko method [3.10] is essentially a transfer of the Saguet/Pic mesh from electromagnetics to diffusion problems.

Wong and Wong [3.11] have developed a technique which allows the density of the mesh to be doubled within certain regions of a problem where more detail is required. Figure 3.19 shows a mesh of length $\Delta x/2$ embedded within a mesh of length Δx. On the right-hand side of the figure different types of nodes are identified: some are common to both meshes, some are associated with one or other mesh and there is also a set of interface nodes. The steps are as follows:

1. Perform a standard TLM routine on the coarse mesh
2. Determine the interface node voltages by finite difference interpolation

$$V_{f0} = \frac{V_{C1} + V_{C2} + V_{C3} + V_{C4}}{4} \tag{3.40}$$

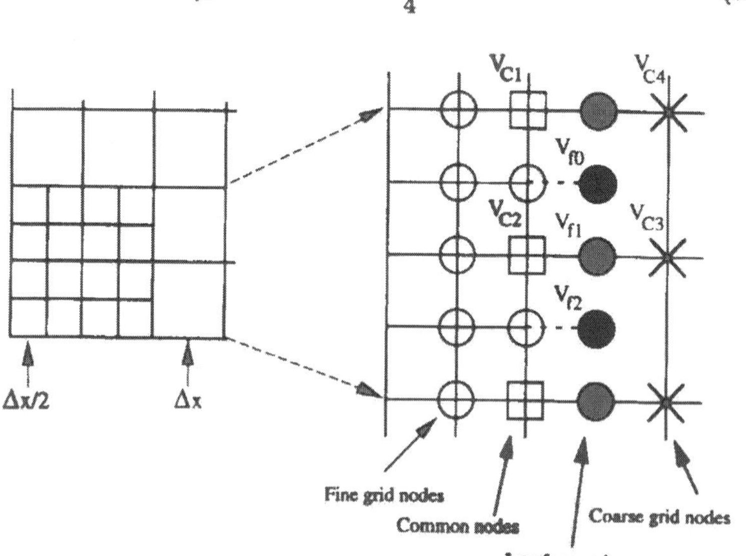

Figure 3.19 A mesh of length $\Delta x/2$ embedded within a mesh of length Δx with the nodes near the interface identified. Note that it is possible to have multiple levels of mesh bisection within a single TLM routine.

$$V_{f1} = \frac{V_{C2} + V_{f0} + V_{C3} + V_{f2}}{4}$$

3. Determine the boundary voltages in the fine mesh using the interface voltages as boundary conditions.
4. Transfer fine mesh voltages back to the coarse mesh. Since the node voltages on the coarse mesh have been modified the scattered (or incident) voltages must also be altered.

Whereas the previous method has two independent meshes with interconnection, Ait-Sadi *et al* [3.12] have developed an alternative approach which maintains synchronism throughout the entire system. The 'dangling' lengths of transmission line in the fine mesh of figure 3.19 are connected to the coarse mesh by means of *bus-bars* (see figure 3.20), a concept borrowed from power electrical engineering. These are placed half way along each line so that nodal incidence and scattering can proceed as if each mesh was independent of the other. The difference lies in the method of connection. In each case we must consider the scattered pulses and what they would see as they arrive at the junction. If we take ${}^{s}V_1(x, y)$, it is travelling along a line of impedance Z_1 but sees impedance Z_1 and Z_2 in parallel as a load.

The incident pulses are then

$$\begin{aligned} {}_{k+1}^{\;\;i}V_1(x, y) &= \frac{2Z_1}{Z_1 + 2Z_2}\, {}_k^sV_2(x', y') + \frac{2Z_2}{Z_1 + 2Z_2}\, {}_k^sV_1(x, y + 1) \\ &\quad - \frac{Z_1}{Z_1 + 2Z_2}\, {}_k^sV_1(x, y + 1) \end{aligned}$$

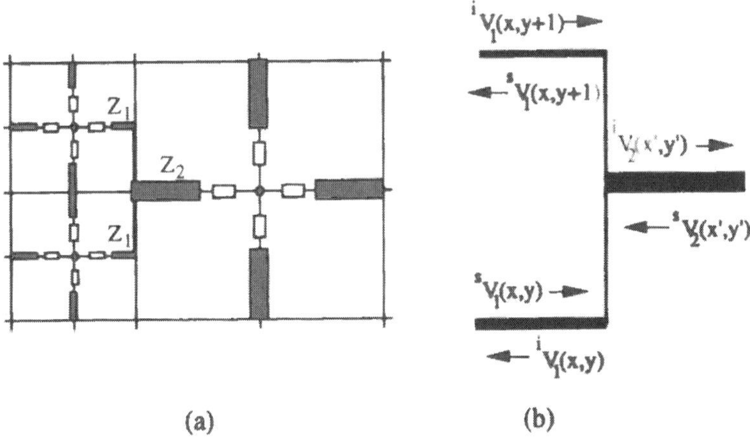

Figure 3.20 (a) The connection of a coarse and fine mesh by means of bus-bars. (b) Scattering in the region of a bus-bar.

$$\begin{aligned}
{}_{k+1}^{\,i}V_1(x, y+1) &= \frac{2Z_1}{Z_1 + 2Z_2}\, {}_k^r V_2(x', y') + \frac{2Z_2}{Z_1 + 2Z_2}\, {}_k^r V_1(x, y) \\
&\quad - \frac{Z_1}{Z_1 + 2Z_2}\, {}_k^r V_1(x, y+1) \\
{}_{k+1}^{\,i}V_2(x', y') &= \frac{Z_1 - 2Z_2}{Z_1 + 2Z_2}\, {}_k^r V_2(x', y') + \frac{2Z_2}{Z_1 + 2Z_2}\, [{}_k^r V_1(x, y) \\
&\quad + {}_k^r V_1(x, y+1)]
\end{aligned} \tag{3.41}$$

Pulko *et al* [3.13] have developed a very similar algorithm involving the use of discontinuous lines to substructure a region of space. From this viewpoint it is clear that the approach is not too removed from the one-dimensional link-resistor formulations. Summation and scattering are as usual in link-line nodes. However there is a secondary node located at the interface between the two meshes as shown in figure 3.21. Pulses pass by at the half time-steps and the potential at the intermediate node is then:

$$\phi_{\text{intermediate}} = \frac{\left[\dfrac{2\,{}^sV_A}{Z_2} + \dfrac{2\,{}^sV_{B1} + 2\,{}^sV_{B2}}{Z_1}\right]}{\left[\dfrac{1}{Z_2} + \dfrac{2}{Z_1}\right]} \tag{3.42}$$

This interface is in fact an additional scattering zone so that the incident pulses are:

$$\begin{aligned}
{}_{k+1}^{\,i}V_A &= \phi_{\text{intermediate}} - {}_k^s V_A \\
{}_{k+1}^{\,i}V_{B1} &= \phi_{\text{intermediate}} - {}_k^s V_{B1} \\
{}_{k+1}^{\,i}V_{B2} &= \phi_{\text{intermediate}} - {}_k^s V_{B2}
\end{aligned} \tag{3.43}$$

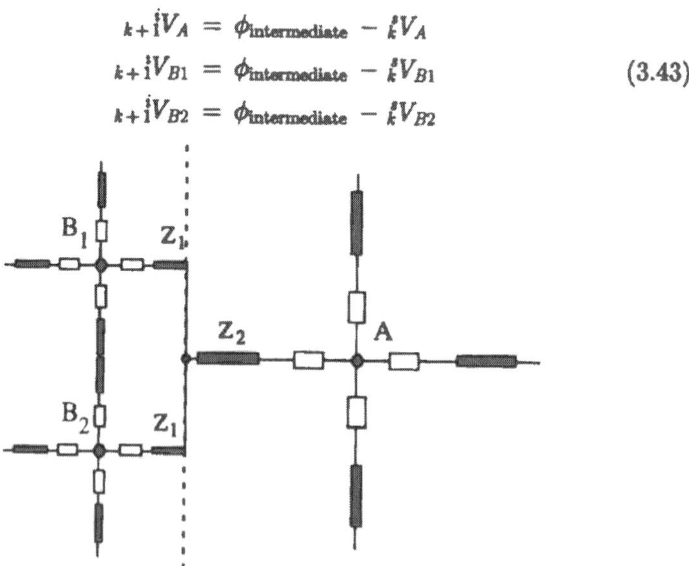

Figure 3.21 The intermediate scattering zone in the Pulko spatial sub-structuring technique.

This section has outlined a series of techniques for dealing with irregular meshes. Up to now there has been no investigations on the possible use of conformal mapping techniques to transform a thermal/diffusion problem from an irregular geometrical space. We have also omitted multi-branch and non-orthogonal meshes such as those which Witwit [3.14] has developed from the electromagnetic work of Simons and Sebak [3.15] and Meliani [3.16].

Time-step variation

There are many applications where the efficiency of the TLM routine is sacrificed on account of some external factor which dictates the magnitude of the time-step. This frequently occurs when an inhomogeneous problem involves flux between materials of significantly different diffusion parameters. In terms of heat-flow this might mean that a poor thermal conductor (e.g. water) is in contact with a good thermal conductor (e.g. metal). If we were to use a value of Δt which was totally appropriate to water, then we would not accurately model the detailed space-time history of the metal. If the situation were reversed then the very much smaller time-step appropriate to the metal would mean that we were simulating little or no change in the water. Johns (in private conversations with the author) pointed out that within a short period of time the metal would approach thermal equilibrium and its subsequent time-history would be determined by the water. He suggested that it should be possible to start with a time-step for metal and as time progressed this could be altered until the simulation was operating at the much longer time-step. He developed a technique which published posthumously by Pulko [3.17]. This considered the scattered pulses as they pass through the mid-point of the link transmission. Thus V_A and V_B of figure 3.22 become V'_A and V'_B after the time-step transformation. The technique is based on two fundamental assumptions:

(1) Total voltage is conserved through the transformation

$$V_A + V_B = V'_A + V'_B \tag{3.45}$$

(2) Total current is conserved through the transformation

$$\frac{V_A - V_B}{Z} = \frac{V'_A - V'_B}{Z'} \tag{3.46}$$

These two equations can then be used to derive values for the transformed pulses:

$$\begin{bmatrix} V'_A \\ V'_B \end{bmatrix} = \frac{1}{2} \begin{bmatrix} (1 + \frac{Z'}{Z}) & (1 - \frac{Z'}{Z}) \\ (1 - \frac{Z'}{Z}) & (1 + \frac{Z'}{Z}) \end{bmatrix} \begin{bmatrix} V_A \\ V_B \end{bmatrix} \tag{3.47}$$

This approach (as it stands) requires that there is no impedance mismatch on the link-lines where the impedance transformation is to take place. In this context it is interesting to note the comment by Pulko [3.13] in relation to spatially sub-structuring where two regions are being modelled at different time-steps. If the difference in link-line impedance is solely due to the difference in time-steps then pulses may be exchanged without modification.

Webb and Gui [3.18] have proposed an alternative time transformation which takes account of any impedance mismatch which might be present due to extraneous factors such as non-linearities in the scattering parameters which are covered in the next section.

The pulse which is incident from the left at node x after time transformation and impedance mismatch reflection is:

$${}_{k+1}^{\;\;i}V_L(x) = \rho'_L(x)\,{}_k^sV_L(x) + \tau'_R(x-1)\,{}_k^sV_R(x-1) \tag{3.48}$$

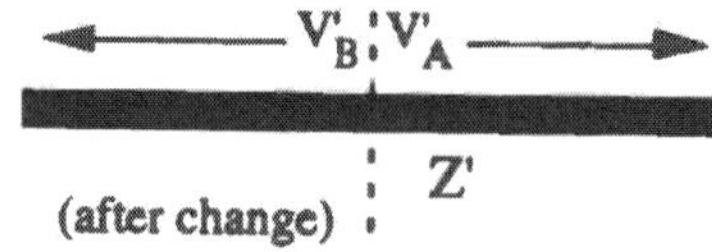

Figure 3.22 Time-step transformation when the line impedance changes from Z to Z'.

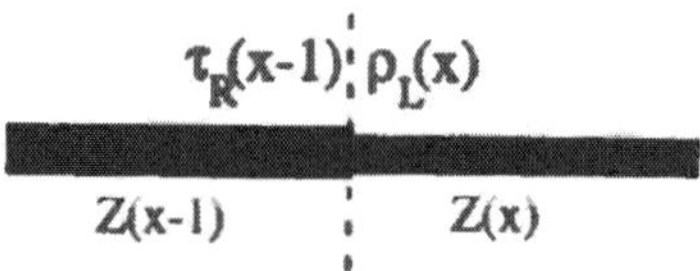

Figure 3.23 Adjacent lines in the Webb/Gui time transformation.

The reflection and transmission coefficients are defined *after* the time transformation and have to be determined.

We define

$$\alpha = \frac{\Delta t'}{\Delta t} = \frac{Z'(x)}{Z(x)} = \frac{Z'(x-1)}{Z(x-1)}$$

We will also define the transmitted and reflected voltages across the mismatch as ${}^{T}V$ and ${}^{R}V$.

Voltage conservation at the mismatch gives:

$$V = \frac{{}^{T}V}{\alpha} - \frac{{}^{R}V}{\alpha} \tag{3.49}$$

Current conservation gives

$$\frac{V}{Z'(x)} = \frac{{}^{T}V}{Z'(x-1)} + \frac{{}^{R}V}{Z'(x)} \tag{3.50}$$

We then get

$$\frac{\tau_L'(x)}{\alpha} - \frac{\rho_L'(x)}{\alpha} = 1 \tag{3.51}$$

and

$$\frac{\tau_L'(x)}{Z(x-1)} + \frac{\rho_L'(x)}{Z(x)} = \frac{1}{Z(x)} \tag{3.52}$$

Therefore

$$\tau_L'(x) = \frac{(1+\alpha)Z(x-1)}{Z(x-1)+Z(x)} \tag{3.53}$$

and

$$\rho_L'(x) = \frac{Z(1-x) - \alpha Z(x)}{Z(x-1)+Z(x)} \tag{3.53}$$

These can be checked for consistency:
if $Z(x) = Z(x-1)$ then the results are identical with the Pulko/Johns [3.17] method. If $Z(x) \neq Z(x-1)$, but $\alpha = 1$ then

$$\rho_L'(x) = \rho_L(x) = \frac{Z(x-1) - \alpha Z(x)}{Z(x-1)+Z(x)}$$

and

$$\tau'_L(x) = \tau_L(x) = \frac{2Z(x-1)}{Z(x-1) + Z(x)}$$

Since voltage and current are conserved, Webb and Gui point out that this is effectively the conservation of power:

$$\frac{V^2}{Z(x)} = [{}^TV - {}^RV]\left[\frac{{}^TV}{Z'(x-1)} + \frac{{}^RV}{Z'(x)}\right] \tag{3.54}$$

Non-linearities in material properties

Because lossy TLM algorithms do not have the same stability criteria as apply in explicit finite difference formulations it is easier to incorporate material non-linearities. In most situations these are due to the influence of the nodal potential (or its analogues) on the magnitude of the local resistance and/or capacitance. The first question that should be asked is whether the non-linearity affects the impedances or the resistances in a node as this will have an important bearing on the subsequent development of an efficient algorithm. If the non-linearities manifest themselves entirely in the nodal resistance then there are few problems. However, the situation is quite different if the capacitance changes as a function of the local nodal potential: a network of initially matched transmission lines can quickly develop inter-nodal mismatches as the simulation proceeds. This of course depends on whether a link-line or a link-resistor formulation is used.

de Cogan and Johns [3.19] were amongst the first to use the direct method for the incorporation of temperature dependent thermal parameters. The arrival of pulses at nodes was assumed to represent the end of an iteration. The values of ${}_k\phi(x)$ were then used to update the model parameters (thermal conductivity, $k_T(x)$ and specific heat, $C_p(x)$. The thermal resistance and capacitance were re-calculated and the resulting scattering coefficients $\rho(k_T({}_k\phi(x)), C_p({}_k\phi(x)))$ and $\tau(k_T({}_k\phi(x)), C_p({}_k\phi(x)))$ were used to calculate the nodal scattered voltages then internodal scattering and the incident voltages at the next time-step.

There are several ways in which information on the dependent parameters can be incorporated into a TLM algorithm and these will be discussed in relation to applications in chapter 5. One approach involves the interpolation of a data-set. In the case of thermal modelling the temperature dependence of specific heat and thermal conductivity is available for a wide range of materials (see for instance [3.20, 3.21]). Henini [3.22–3.24] included in his algorithms a set of data for all materials within a problem. At the end of each time-step an interpolation routine used the appropriate data-set to calculate the thermal parameters for the next iteration. Interpolation can be avoided and computational load can be reduced if the thermal data for a material is expressible in a polynomial form. For example, the specific heat as a function of temperature for metals can be given in the form [3.25]:

$$C_p = a + (b \times 10^{-3})T + (c \times 10^{-6})T^2 + \frac{d \times 10^5}{T^2} \qquad (3.55)$$

a, b, c and d are constants for a given material and T is the temperature (in Kelvin).

Ongoing changes in thermal conductivity do not cause a significant problem, but substantial changes in specific heat certainly do. Changes in thermal capacitance means that the line impedances are altered and thus the reflection coefficients must be recalculated for the entire mesh at every iteration. In the case of link-line algorithms there may be an unacceptable increase in the computational load. An alternative approach which avoids internodal reflections is to incorporate changes in capacitances into stubs and ensure that the link-lines are matched at all times [3.26]. If a model is based on a link-resistor nodal configuration (as in [3.22]) then the problem of internodal reflections is avoided; we are effectively using the values of ${}_k\phi(x, y, z)$ to calculate the nodal scattering coefficients that are used at the next time-step.

de Cogan and Henini [3.27] noted that in certain situations the extreme nature of a non-linearity may make it inappropriate for $C({}_k\phi(x, y, z))$ and $R({}_k\phi(x, y, z))$ to be used to calculate ${}_{k+1}\phi(x, y, z)$, rather $C({}_{k+1}\phi(x, y, z))$ and $R({}_{k+1}\phi(x, y, z))$ should be used. At first sight this appears to change TLM from an explicit to an implicit routine. However, it has been found that an iteration-within-iteration works very well. At the start of a new time-step $C({}_{k+1}\phi(x, y, z))$ and $R({}_{k+1}\phi(x, y, z))$ are used to make a first estimate of ${}_{k+1}\phi(x, y, z)$. This is then used to make a first estimate of $C({}_{k+1}\phi(x, y, z))$ and $R({}_{k+1}\phi(x, y, z))$. These in turn are used with the incident pulses at the k-th iteration to make a second estimate of ${}_{k+1}\phi(x, y, z)$. The steps are repeated until convergence is achieved which is rarely more than about 3–4 iterations.

It might indeed be possible to improve on existing algorithms by recognising that internodal scattering, when it exists, occurs at the half-times (*e.g.* $(k+1)/2$). Independent of the nodal configuration it should not be difficult to use the techniques which have been outlined above to calculate the potentials at this instant. Thus $C({}_k\phi(x))$ and $R(({}_k\phi(x))$ would determine what is scattered from the nodes while $C({}_{(k+1)/2}\phi(x \pm 1/2))$ and $R({}_{(k+1)/2}\phi(x \pm 1/2))$ would determine what is incident at $k+1$.

References

[3.1] P. Enders, D. de Cogan and A. Soulos, TLM: order of accuracy enhancement, *Second International Workshop on Discrete Time Domain, Modelling of Electromagnetic Fields and Networks*, Hotel Ambassador, Berlin 28, 29 October 1993.

[3.2] D.A. Al-Mukhtar and J.E. Sitch, Transmission line matrix method with irregularly graded space, *IEE Proc. (H)*, **128** (1981) 299–305.

[3.3] P. Naylor, Variable mesh size and circular TLM, *MSc thesis*, Nottingham University 1982..

[3.4] D. de Cogan and S.A. John, A Two Dimensional TLM Model for the Punch-Through Diode, *J. Phys. D.*, **18** (1985) 507–515.

[3.5] R. Cacoveanu, P. Saguet and F. Ndagijimana, TLM method: a new approach for the central node in polar meshes, *Electronics Letters*, **31** (1995) 297–298.

[3.6] D. de Cogan, A.K. Shah and M. Henini, Variable Mesh TLM Modelling of Heat Flow in Semiconductors, *Proc. of Fourth Int. Conf. on Numerical Analysis of Semiconductor Devices* (NASECODE IV), Dublin 1985, p. 255.

[3.7] D. de Cogan and A.K. Shah, Stub TLM Modelling of Heat Flow in Semiconductors, *J. Phys. D.*, **19** (1986) 721–715.

[3.8] P. Saguet and E. Pic, Le maillage rectangulaire et le changement de maille dans la methode TLM en deux dimensions, *Electronics Letters*, **17** (1981) 277–279.

[3.9] D. de Cogan, P. Enders and X. Gui, Impedance transformations and mesh coarsening in TLM heat-flow modelling, *Numerical Heat Transfer*, **21**(B) (1992) 327–342.

[3.10] S.H. Pulko, A. Mallik and P.B. Johns, Application of transmission line modelling (TLM) to thermal diffusion of bodies of complex geometry, *Int. Jnl. Numerical Methods in Engineering*, **23** (1986) 2303–2313.

[3.11] C.C. Wong and W.S. Wong, Multigrid TLM for diffusion problems, *International Journal of Numerical Modelling*, **2** (1989) 103–111.

[3.12] R. Ait-Sadi, A.J. Lowery and B. Tuck, Combined coarse-fine transmission line modelling method for diffusion problems, *International Journal of Numerical Modelling*, **3** (1990) 111–126.

[3.13] S.H. Pulko, J.A. Halleron and C.F. Phizacklea, Substructuring of space and time in TLM diffusion applications, *International Journal of Numerical Modelling*, **3** (1990) 207–214.

[3.14] A.R.M. Witwit, Meshing techniques for TLM diffusion problems, *PhD thesis*, Hull University (UK) 1995.

[3.15] N.R.S. Simons and A.A. Sebak, Spatially weighted numerical models for the two-dimensional wave equation: FD algorithm and synthesis of the equivalent TLM model, *International Journal of Numerical Modelling*, **5** (1992) 111–130.

[3.16] H. Meliani Mesh generation in TLM, *PhD thesis*, Nottingham University 1987.

[3.17] S.H. Pulko, A. Mallik, R. Allen and P.B. Johns, Automatic timestepping in TLM routines for the modelling of thermal diffusion processes, *International Journal of Numerical Modelling*, **3** (1990) 127–136.

[3.18] P.W. Webb and X. Gui, Time-step changes in TLM diffusion modelling, *International Journal of Numerical Modelling*, **5** (1992) 251–257.

[3.19] D. de Cogan and S.A. John, The calculation of temperature distribution in punch-through structures during pulsed operation using the transmission line matrix (TLM) method, *J. Phys D*, (Applied Physics), **15** (1982) 1979–1990.

[3.20] Y.S. Touloukian, Thermophysical Properties of Materials Vol. 5 (Specific Heat) IFI Plenum, New York 1972.

[3.21] Y.S. Touloukian, Thermophysical Properties of Materials Vol. 2 (Thermal conductivity) IFI Plenum, New York 1972.

[3.22] D. de Cogan and M. Henini, 3D-TLM Modelling of the thermal behaviour of conducting films on insulating substrates, *J. Phys. D.*, 20 (1987) 1445 – 1450.

[3.23] D. de Cogan and M. Henini, TLM Modelling of solder voids in power semiconductors *IEEE Proc. Components, Hybrids and Manufacturing Technology* CHMT-10 (1987) 440 – 445.

[3.24] S.V.G. Vardigans, D. de Cogan and M. Henini, A transient thermal failure mode in metal film resistors, *J. Phys. D.*, 20 (1987) 1454 – 1456.

[3.25] Handbook of Chemistry and Physics (64th Edition), CRC Press 1983, p. D43.

[3.26] S.H. Pulko and P.B. Johns, Modelling of thermal diffusion in three dimensions by the transmission line matrix method and the incorporation of non-linear thermal properties, *Comms. in Applied Numerical Methods*, 3 (1987) 571 – 579.

[3.27] D. de Cogan and M. Henini, Thermal effects in thin conducting films on silica and alumina substrates under near adiabatic conditions, *J. Phys. D.*, 20 (1987) 1451 – 1453.

Chapter 4

The theory of lossy TLM

Introduction

Lossless TLM has a firm theoretical basis and has a niche applications area in which it is superior to other techniques. On the other hand TLM for diffusion modelling has had 'psychological' problems. These are largely due to questions of its identity, its pedigree and its utility. Conventional lossy treatments start with the Telegraphers' equation and immediately ignore the existence of a major component while trying to treat problems that have long been the realm of finite difference methods.

In this chapter we will be addressing questions such as:

- What is the basis for lossy TLM?
- Are assumptions valid for all normal circumstances?
- Is the technique accurate?
- Is the technique consistent?
- How does it compare with other techniques?

We do not claim to have all the answers for every situation, but we will attempt to provide answers to these questions using a range of different approaches. In the first instance lossy TLM is analysed using the Johns electromagnetic-based method; we will assume that the reader is already conversant with that theory which comes directly from, and is common with, lossless TLM. This is followed by comparisons with finite difference and the state-space equations. TLM is then analysed in terms of probability and statistics. In some cases the treatments may overlap, but it is hoped that this sequence of parallel interpretations will help the reader to make an individual assessment of lossy TLM within the wider context of numerical methods.

An electromagnetic view of lossy TLM

The Johns paper outlining the application of TLM to diffusion [4.1] is in fact two papers rolled into one. The first part lays the groundwork and discusses the relationship with FD and indicates how various FD configurations can be expressed as TL networks. The second part is concerned with '*Propagation Analysis*' but is presented in such a concise way that the utilisation of his concepts is not easy. For that reason we include a comprehensive exposition of the technique in this chapter.

In the solution of differential equations it is normal to investigate the stability by means of frequency analysis and this is done in the section dealing with the comparison with finite difference. Johns could have done the same here, but instead he choose to develop his own technique which has recently been revisited by Kenny [4.2]. The John's technique (*Propagation Analysis*) is based on a combination of electromagnetics, electrical network theory and electrical filter theory. It takes a plane wave of frequency, f, which interacts with the mesh and sees how the interaction is influenced by spatial and temporal discretisations.

The relationship between the input and output of an electrical network can be expressed in terms of 'a' parameters:

$$\begin{bmatrix} V_{IN} \\ I_{IN} \end{bmatrix} = \mathbf{A} \begin{bmatrix} V_{OUT} \\ I_{OUT} \end{bmatrix} \quad \text{where } \mathbf{A} = \begin{bmatrix} a_{11} & a_{12} \\ a_{21} & a_{22} \end{bmatrix} \tag{4.1}$$

The propagation analysis method treats the lossy node as a T-network with series components Z_1 and Z_3 and shunt component Z_2 (see figure 4.1) and derives the a-parameter matrix as:

$$\mathbf{A} = \begin{bmatrix} \dfrac{Z_1 + Z_2}{Z_2} & \dfrac{(Z_1 + Z_2)(Z_2 + Z_3) - Z_2^2}{Z_2} \\ \dfrac{1}{Z_2} & \dfrac{Z_1 + Z_2}{Z_2} \end{bmatrix} \tag{4.2}$$

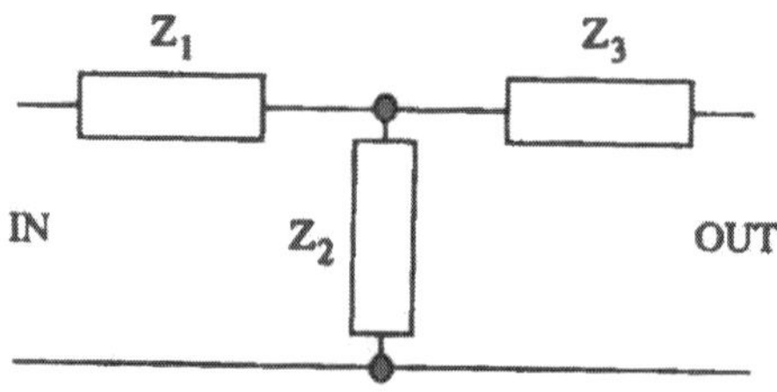

Figure 4.1 A two-port T-network.

In the case of a one-dimensional link-line TLM node there are in effect three separate electrical networks in series. Reading from left to right there is a transmission line, followed by two resistors followed by a transmission line. Each of these can be treated as a T-network and the resultant of this cascade of networks can be expressed as a matrix product [4.3]:

$$\begin{bmatrix} V_{IN} \\ I_{IN} \end{bmatrix} = \mathbf{A}_{\text{line}}\,\mathbf{A}_{\text{resistor}}\,\mathbf{A}_{\text{line}} \begin{bmatrix} V_{OUT} \\ I_{OUT} \end{bmatrix} \tag{4.3}$$

The line matrix is derived from electromagnetics and for a line of impedance Z and length $\Delta x/2$ is given by:

$$\mathbf{A}_{\text{line}} = \begin{bmatrix} \cos(\omega t/2) & Z\,j\sin(\omega t/2) \\ \dfrac{j\sin(\omega t/2)}{Z} & \cos(\omega t/2) \end{bmatrix} \tag{4.4}$$

In the case of the resistor network $Z_1 = Z_3 = R$ and $Z_2 = \infty$.

Therefore

$$\mathbf{A}_{\text{resistor}} = \begin{bmatrix} 1 & 2R \\ 0 & 1 \end{bmatrix} \tag{4.5}$$

In working towards this stage we have in fact already moved to consider the interaction between a sinusoidal wave with $\omega = 2\pi f$ and the node consisting of two half-length transmission lines separated by two resistors. Since we are in the time-domain $\omega t/2$ is replaced by $\theta/2$ where $\theta = 2\pi\Delta x/\lambda$ (Δx is the spatial discretisation and λ is the wavelength of the incident wave).

Before proceeding with the concepts of propagation analysis we should first consider other nodal arrangements:

A link-line node with an open-circuit half-length stub at its centre can be broken into three sections comprising line, centre and line. The centre has $Z_1 = Z_3 = R$. The shunt component, Z_2 is effectively the apparent impedance of a transmission line (length $\Delta x/2$) terminated in an open circ 1.4] (i.e. $Z/(j\tan\theta/2)$) so that the matrix for the centre is:

$$\mathbf{A}_{\text{centre}} = \begin{bmatrix} \frac{R}{Z} j\tan\theta/2 + 1 & 2R + \frac{R^2}{Z} j\tan\theta/2 \\ \frac{1}{R} j\tan\theta/2 & \frac{R}{Z} j\tan\theta/2 + 1 \end{bmatrix} \tag{4.6}$$

If we had a lossy stub (i.e. a half-length open-circuit stub in series with a resistor then this effectively shunts the signal so that $Z_2 = R + Z/(j\tan\theta/2)$.

In the case of a two-dimensional link-line node then there are two lines at right angles to the propagating wave with a total impedance of $(R+Z)/2$.

The line impedance is Z because within the time-step Δt each line appears to be infinitely long and $Z_2=(R+Z)/2$.

The total a-matrix for a link-resistor node is

$$\mathbf{A}_{\text{resistor}}\,\mathbf{A}_{\text{line}}\,\mathbf{A}_{\text{resistor}} \tag{4.7}$$

where

$$\mathbf{A}_{\text{line}} = \begin{bmatrix} \cos\theta & jZ\sin\theta \\ j\sin\theta/Z & \cos\theta \end{bmatrix} \quad \text{and} \quad \mathbf{A}_{\text{resistor}} = \begin{bmatrix} 1 & R \\ 0 & 1 \end{bmatrix} \tag{4.7}$$

On this occasion we have θ and not $\theta/2$ because the length of the line within the node is Δx.

In the case of any of these nodal configurations the total transmission matrix A can be used to find an expression for the wave propagation constant, γ because $A_{11}=\cosh(\gamma\Delta x)$.

This is useful because the wave propagation constant can be expressed as $\gamma=\alpha+j\beta$ where α is the attenuation which a wave experiences as it passes through the node and β is the resulting phase-shift. Johns [4.1] demonstrated that diffusion could be modelled by a simple link-line node ($R_2=\infty$) where A_{11} is:

$$\cosh(\gamma\Delta x) = \cos\theta + \frac{j}{2k}\sin\theta \tag{4.8}$$

in this case $k = D\Delta t/\Delta x^2$ (D is diffusion constant)
This can now be expressed as:

$$\cosh(\alpha + j\beta) = u + jv \tag{4.9}$$

and the equation is then solved in terms of α and β.

The solution of this equation is not readily available in text-books but it can be approached as follows:

$$\alpha = \cosh^{-1}(s)$$

$$\beta = \cos^{-1}(t)$$

where

$$s = \frac{r_1 + r_2}{2} \qquad t = \frac{r_1 - r_2}{2}$$

r_1 and r_2 are given by:

$$r_1 = \sqrt{(1+u)^2 + v^2} \qquad r_2 = \sqrt{(1-u)^2 + v^2}$$

These are multi-valued functions and there can be problems which will be discussed presently. The tables for this equation by Kennelly [4.5] include instructions on how to ensure a correct interpretation.

The case of the simple link-line node which is described by $\cosh(\gamma\Delta x) = \cos\theta + (j\sin\theta)/2k$ has $u = \cos\theta$ and $v = (\sin\theta)/2k$. The resulting solutions for α and β can be plotted as a function of θ (sinusoidal wave number for constant discretisation, Δx) and these are shown in figure 4.2(a). A parametric plot of β versus α is shown in figure 4.2(b). This demonstrates how the attenuation, (α) deviates from the analytical result (dashed line in the figure) as the phase (β) is changed. Johns used the technique to show how this simple TLM representation compared with the equivalent explicit and implicit finite difference routines as a function of λ and diffusion number. He showed that at $k=0.5$ the link-line node and the explicit finite difference routine (at its stability limit) were identical.

The propagation curves for a variety of nodal configurations were studied in detail by Moravec [4.6]. She also examined the equivalent behaviour of the network corresponding to an explicit finite difference routine where the

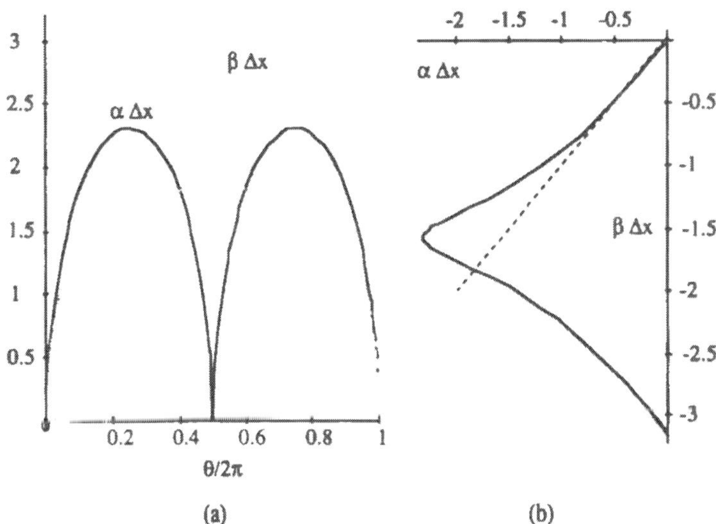

Figure 4.2 (a) α and β vs. θ, (b) b vs. α parametric plot for a one-dimensional link-line node with $k=0.1$. The dashed line in (b) is the analytical result.

u and v terms describing $\cosh(\gamma\Delta x)$ are $(1-\theta^2/4k)$ and $\theta/2k$ respectively. A MATLAB plot which ignores the multiplicity of the inverse functions is shown in figure 4.3(a) and the corrected plot is shown in figure 4.3(b). From this we can clearly see that instability occurs in a finite difference node when there are frequency components for which the attenuation constant becomes positive (amplifies) within the region of phase lag $(0 \leq \beta \leq \pi)$.

When the same analysis is completed for a link-resistor node then the results are similar to eqn (4.8) except that the v term is twice as large (i.e. $\sin(\theta)/k$) and this means that for the one-dimensional case the $\alpha\Delta x$ vs. $\beta\Delta x$ curve is closer to the analytical result over a larger frequency range than the link-line node or indeed either the explicit or implicit finite difference techniques.

Propagation analysis is without doubt a very powerful technique, but because its implementation requires a broad range of experience in electromagnetics and complex functions it has not been widely used. Nevertheless it does demonstrate the range of space and time discretisations as well as frequencies which give similar results and/or provide an accurate approximation to the analytical result. There is much scope for investigations such as the behaviour of two- and three-dimensional link-line and link-resistor networks with and without a range of stub configurations (Peter Johns, in private discussions with the author, stated that he believed that a series combination of stub resistor and open-circuit line could give a closer approximation to the analytical equations for diffusion).

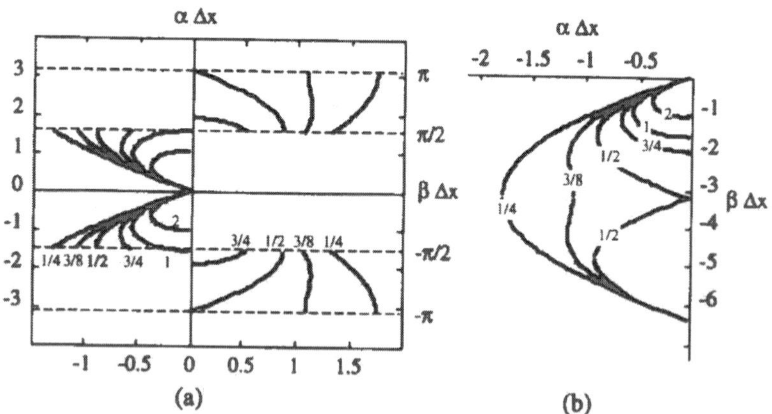

Figure 4.3 β vs. α plots as a function of diffusion number for the explicit finite difference routine. (a) as calculated using MATLAB (b) following correction for inverse function ambiguity.

TLM as a finite difference approximation

In finite difference schemes the partial derivatives $\partial V/\partial x$ can be expressed in several ways:

$\dfrac{V(x+1)-V(x)}{\Delta x}$ forward difference (in space) approximation of $\partial V/\partial x$ with order of 1 error $O(\Delta x)$

$\dfrac{V(x)-V(x-1)}{\Delta x}$ backward difference with order of error $O(\Delta x)$

$\dfrac{V(x+1)-V(x-1)}{2\Delta x}$ central difference with order of error $O(\Delta x)^2$

$\dfrac{V(x+1)+V(x+1)-2V(x)}{\Delta x^2}$ central difference approximation of $\dfrac{\partial^2 V}{\partial x^2}$ with order or error $O(\Delta x)^2$

Similar approximations can be made for time derivatives and the diffusion equation can be represented by combinations of these finite differences. A scheme which uses central differences for space and time will not work because it will always be unstable. On the other hand the following expression

$$\frac{1}{\Delta t}[{}_{k+1}V(x) - {}_kV(x)] = \frac{1}{D\Delta x^2}[{}_kV(x+1) + {}_kV(x-1) - 2\,{}_kV(x)]$$

or

$$_{k+1}V(x) = {}_kV(x) + \frac{\Delta t}{D\Delta x^2}[{}_kV(x+1) + {}_kV(x-1) - 2\,{}_kV(x)] \qquad (4.10)$$

(normally called the *explicit* finite difference scheme) is stable so long as the factor $\Delta t/(D\Delta x^2) \leq 0.5$. It is one-step because it only requires ${}_kV$ to calculate ${}_{k+1}V$. It is explicit because ${}_{k+1}V$ is calculated entirely in terms of ${}_kV$ values.

An alternative scheme which is due to duFort and Frankel [4.7] replaces the 2 ${}_kV(x)$ term in the central difference approximation of $\partial^2 V/\partial x^2$ by a mean value *in time*. The total approximation for the diffusion equation is:

$$ {}_{k+1}V(x) = \frac{D\Delta x^2}{2\Delta t + D\Delta x^2}\, {}_{k-1}V(x) + \frac{2\Delta t}{2\Delta t + D\Delta x^2}\left[{}_k V(x+1) + {}_k V(x-1)\right] \tag{4.11} $$

This is explicit and unconditionally stable, but now uses ${}_{k-1}V$ and ${}_kV$ to calculate ${}_{k+1}V$; it is two-step.

Whereas TLM adopted the propagational analysis of Johns [4.1] as a means of confirming unconditional stability the conventional spectral analysis technique for the determination of the stability of a finite difference scheme starts by looking at the propagation of a single initial condition considered as a component in a Fourier series.

$$ {}_0V(x) = e^{j\beta x\Delta x} \tag{4.12} $$

As this propagates the mesh value at point x and time k can be expressed as:

$$ {}_kV(x) = \xi^k e^{j\beta x\Delta x} \tag{4.13} $$

ξ is the amplification factor and is a complex function which depends on k and β. It must be less than unity for a system not to go unstable.

In order to determine the stability of a particular finite difference scheme the expression in eqn (4.13) is modified and substituted for each term. After algebraic manipulation we can then find the conditions which gives $\xi \leq 1$.

Finite difference modellers are also concerned with *consistency*. It is possible to have a stable scheme for the diffusion equation that converges to a result that is different from the analytical result as Δx and Δt are reduced towards zero. We would then be forced to say that the scheme was *not consistent* with the diffusion equation.

If we were to take the explicit finite difference equation (4.10) and compare it with the equivalent analytical solution we would find that they were not the same. This is due to the *truncation error* and is given by:

$$ \frac{1}{2}\,\Delta t\,\frac{\partial^2 V}{\partial t^2} + \frac{1}{6}\,\Delta x^2\,\frac{\partial^4 V}{\partial x^4} \tag{4.14} $$

However, as Δt and $\Delta x \to 0$, then so does this error.

In the case of the duFort-Frankel scheme the truncation error is:

$$ D\frac{\Delta t^2}{\Delta x^2}\,\frac{\partial^2 V}{\partial t^2} + \frac{1}{6}\,\Delta t^2\,\frac{\partial^3 V}{\partial t^3} + \frac{1}{4}\,D\Delta x^2\,\frac{\partial^4 V}{\partial x^4} \tag{4.15} $$

Now, as Δt and $\Delta x \to 0$ the error tends towards $D(\Delta t^2/\Delta x^2)\partial^2 V/\partial t^2$.
If Δt and Δx tend to zero at the same rate then the approximation is consistent with an equation of the form

$$\frac{\partial^2 V}{\partial x^2} = A\frac{\partial^2 V}{\partial t^2} + \frac{\partial V}{\partial t} \tag{4.16}$$

This is certainly not the diffusion equation, but is more like the telegraphers' equation of TLM and will be revisited later.

TLM error parameters

In its most general form TLM modellers express the lossy wave equation as:

$$\nabla^2 V = a\,L_d C_d\,\frac{\partial V^2}{\partial t^2} + b\,R_d C_d\,\frac{\partial V}{\partial t} \tag{4.17}$$

Because of the partitioning of the distributed components within the node the constants are $b = 1, 2, 3$ depending on dimension and $a = 2b$.

The simplistic approach is to say that the dispersive term dominates so that:

$$a\,L_d C_d\,\frac{\partial V^2}{\partial t^2} \ll b\,R_d C_d\,\frac{\partial V}{\partial t} \tag{4.18}$$

Pulko [4.8] described a parameter, m, which expressed the ratio of these two terms.

$$m = \frac{\left[a\,L_d C_d\,\frac{\partial V^2}{\partial t^2}\right]}{\left[b\,R_d C_d\,\frac{\partial V}{\partial t}\right]} = \frac{\left[a\,Z\Delta t\,\frac{\partial V^2}{\partial t^2}\right]}{\left[b\,R\,\frac{\partial V}{\partial t}\right]} \tag{4.19}$$

The two derivatives are then presented as finite differences to yield

$$m = \frac{a\,Z({}_k\phi - 2\,{}_{k-1}\phi + {}_{k-2}\phi)}{b\,R({}_k\phi - {}_{k-1}\phi)} \tag{4.20}$$

However, in its simplest form this results in numerical oscillations. Pulko attempted to overcome this problem by using mean values over previous time-steps so that:

$${}_k\phi \to \frac{{}_k\phi - {}_{k-1}\phi}{2} \qquad {}_{k-1}\phi \to \frac{{}_{k-2}\phi - {}_{k-3}\phi}{2} \qquad {}_{k-2}\phi \to \frac{{}_{k-4}\phi - {}_{k-5}\phi}{2}$$

Gui *et al* [4.9] investigated the behaviour of the m-parameter. A series of numerical experiments suggested that while it might be a measure of the ratio of the wave and diffusive components it did not necessarily reflect the accuracy of solution. Enders and de Cogan [4.10] suggested an alternative parameter which was intended to reflect both the ratio of the of the two components in the telegraphers' equation and the accuracy of solution. However, no related experiments were undertaken and it does not seem to have been adopted as an accuracy measure.

Pulko [4.8] has suggested in outline an alternative method for quantifying the error due to a significant inductive component in the telegraphers' equation. At the instant of arrival at the end of a transmission line there will be a potential due to the super-position of incident and reflected pulses:

$$\begin{aligned} {}_kV_L(x) &= {}_k^iV_L(x) + {}_k^rV_L(x) \\ {}_kV_R(x) &= {}_k^iV_R(x) + {}_k^rV_R(x) \end{aligned} \tag{4.21}$$

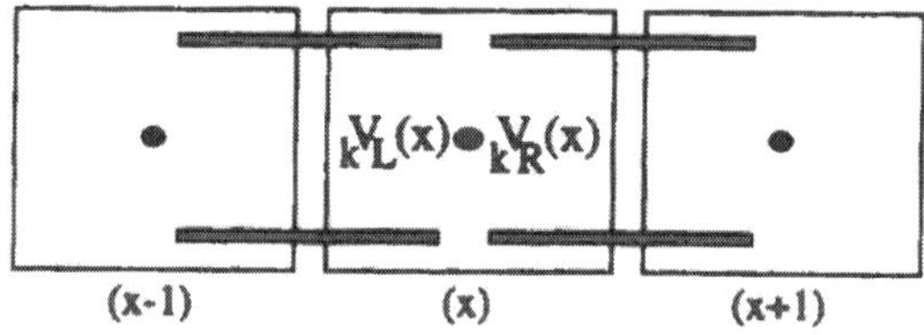

Figure 4.4 Adjacent nodes with link-lines and the components of eqn (4.21).

If the inductive component in each of the lines in figure 4.4 is negligible then we should be able to say that:

$${}_kV_L(x) \approx {}_kV_R(x-1) \text{ and } {}_kV_R(x) \approx {}_kV_L(x+1)$$

However it has been pointed out [4.11, 4.12] that the exact nature of the initial conditions can lead to oscillatory behaviour and it is possible that

$${}_kV_L(x) \rightarrow {}_{k+1}V_R(x-1) \quad {}_{k+1}V_L(x) \rightarrow {}_kV_R(x+1) \text{ and so on}$$

For this reason a mean value of the difference in potentials taken over two consecutive steps is suggested as a method of obtaining information about the extent of inductive error.

Russell and Webb [4.15] have pursued the idea of the voltage drop across the transmission line as a result of a time-variation in current. If the voltage drop across the link-line between nodes x and $x-1$ is ignored then the current flux from the left arriving at node x at time k can be expressed as:

$$ {}_kI_L(x) \approx \frac{{}_k\phi(x) - {}_k\phi(x-1)}{R(x) + R(x-1)} \tag{4.22} $$

The time derivative of the current can be approximated by a backward difference equation as:

$$ \left(\frac{\partial I_L(x)}{\partial t}\right)_k \approx \frac{{}_kI_L(x) - {}_{k-1}I_L(x)}{\Delta t} \tag{4.23} $$

If the inductor associated with node x is $L(x)$, then the voltage drop due to the time-varying current would be:

$$ {}_kV_L(x) \approx L_{(x)}\left(\frac{\partial I_L(x)}{\partial t}\right)_k \approx L_{(x)}\frac{{}_kI_L(x) - {}_{k-1}I_L(x)}{\Delta t} \tag{4.24} $$

Using Ohm's law we can now rewrite this as:

$$ \begin{aligned} {}_kV_L(x) &\approx \frac{L_{(x)}}{\Delta t}\frac{{}_k\phi(x) - {}_k\phi(x-1) - {}_{k-1}\phi(x) + {}_{k-1}\phi(x-1)}{R(x) + R(x-1)} \\ &\approx Z(x)\frac{{}_k\phi(x) - {}_k\phi(x-1) - {}_{k-1}\phi(x) + {}_{k-1}\phi(x-1)}{R(x) + R(x-1)} \end{aligned} \tag{4.25} $$

There is a similar treatment for ${}_kV_R(x)$ and the sum of magnitudes of these two components would give a measure of the overall error voltage at the TLM node.

Soulos *et al* [4.16] investigated the increase in order of accuracy for one-dimensional diffusion treated as over-damped wave propagation. Order of accuracy measures how the discretisation error vanishes when the mesh is refined. Variations in Z do not achieve an improvement and for this reason the authors considered a loaded node with a stub-line and resistor (R_S) in series. The left, right and stub voltages can be expressed using the traditional *scatter* and *connect* routines. Alternatively, the incident pulses can be expressed in terms of the incident pulses at the previous iteration

$$ {}_{k+1}\begin{pmatrix} {}^iV_L \\ {}^iV_R \\ {}^iV_S \end{pmatrix} = \begin{pmatrix} (1-\rho-\tau')\bar{\mathbf{x}} & \rho\bar{\mathbf{x}} & \tau'\bar{\mathbf{x}} \\ \rho\mathbf{x} & (1-\rho-\tau')\mathbf{x} & \tau'\mathbf{x} \\ \tau_S/2 & \tau_S/2 & \rho_S/2 \end{pmatrix} {}_k\begin{pmatrix} {}^iV_L \\ {}^iV_R \\ {}^iV_S \end{pmatrix} \tag{4.26} $$

where ρ is the link-line reflection, ρ_S and τ_S are the stub reflection and transmission; τ'_S refers to the component which arrives from a link-line and is passed into the stub.

$\mathbf{x}$ and $\bar{\mathbf{x}}$ are nodal shift operators such that $\mathbf{x}V(i) = V(i+1)$ and $\bar{\mathbf{x}}V(i) = V(i-1)$.

This matrix equation can be diagonalised by means of the Caley-Hamilton equation which states that a matrix fulfils its own eigenvalue equation. Therefore each component of V obeys the equations:

$$\begin{aligned} {}_{k+3}^{\,i}V_M &= \{\rho_s + (\tau - \tau')\}\,{}_{k+2}^{\,i}V_M + \left\{ \left[(\tau - \tau')\rho_s - \frac{\tau'\tau_s}{2} \right] \mathbf{X} \right. \\ &\quad \left. + (\tau - \tau')^2 - \rho^2 \right\} {}_{k+1}^{\,i}V_M + \{[(\tau - \tau')^2 - \rho^2]\rho_S \\ &\quad + \tau'\tau_s(\rho - \tau + \tau')^2\}\,{}_{k}^{i}V_M \end{aligned} \tag{4.27}$$

Here the subscript, M reads either L, R or S and $\mathbf{X} = \mathbf{x} + \bar{\mathbf{x}}$

It should be noted that the scheme is 3-step with a much larger number of terms entering the calculation. In the continuum limit of one of these equations the zero-order terms cancel and the scheme is consistent with the diffusion equation up to $O(\Delta t^2 + \Delta x^4)$.

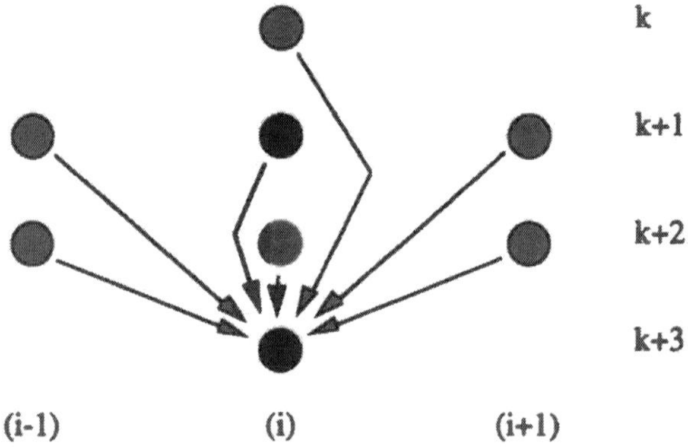

Figure 4.5 The nodal contributions in the high accuracy order TLM scheme of Soulos et al. [4.16].

Kenny [4.2] has taken an entirely different approach to TLM/finite difference comparisons which has some common aspects with Soulos et al [4.16]. He has expressed the nodal potential for a 2-dimensional formulation in very general terms:

$$_{k+1}\phi(x, y) = A \sum_{m=1}^{4} {}_{k+1}^{\,i}V_m(x, y) + B\,{}_{k+1}^{\,i}V_5(x, y) \tag{4.28}$$

where m=1, 2, 3, 4 are the link-line directions; '5' refers to a stub (if present) and A and B are constants.

Scatter and connect are expressed in terms of the potential at adjacent nodes:

$$
\begin{aligned}
{}_{k+1}^{\,i}V_1(x,y) &= {}_k^r V_3(x-1,y) = a_k\phi(x-1,y) + b\,{}_k^i V_3(x-1,y)\\
{}_{k+1}^{\,i}V_2(x,y) &= {}_k^r V_4(x,y+1) = a_k\phi(x,y+1) + b\,{}_k^i V_4(x,y+1)\\
{}_{k+1}^{\,i}V_3(x,y) &= {}_k^r V_1(x+1,y) = a_k\phi(x+1,y) + b\,{}_k^i V_1(x+1,y)\\
{}_{k+1}^{\,i}V_4(x,y) &= {}_k^r V_2(x,y-1) = a_k\phi(x,y-1) + b\,{}_k^i V_2(x,y-1)\\
{}_{k+1}^{\,i}V_5(x,y) &= {}_k^r V_5(x,y) = c_k\phi(x,y) + d\,{}_k^i V_5(x,y)
\end{aligned}
$$

In this case a, b, c and d are constants.

These can be substituted into eqn (4.28) to give

$$
\begin{aligned}
{}_{k+1}\phi(x,y) =\; & Aa[{}_k\phi(x-1,y) + {}_k\phi(x,y+1) + {}_k\phi(x+1,y) + {}_k\phi(x,y-1)]\\
& + Ab[{}_k^i V_3(x-1,y) + {}_k^i V_4(x,y+1) + {}_k^i V_1(x+1,y)\\
& + {}_k^i V_2(x,y-1)] + Bc\,{}_k\phi(x,y) + Bd\,{}_k^i V_5(x,y)
\end{aligned}
\tag{4.30}
$$

Shifting in space and time allows us to attempt a removal of all the incident pulses from this equation.

$$
\begin{aligned}
{}_k^i V_1(x,y) &= {}_{k-1}^{\;r}V_3(x,y) = a\,{}_{k-1}\phi(x,y) + b\,{}_{k-1}^{\;i}V_3(x,y)\\
{}_k^i V_2(x,y) &= {}_{k-1}^{\;r}V_4(x,y) = a\,{}_{k-1}\phi(x,y) + b\,{}_{k-1}^{\;i}V_4(x,y)\\
{}_k^i V_3(x,y) &= {}_{k-1}^{\;r}V_1(x,y) = a\,{}_{k-1}\phi(x,y) + b\,{}_{k-1}^{\;i}V_1(x,y)\\
{}_k^i V_4(x,y) &= {}_{k-1}^{\;r}V_2(x,y) = a\,{}_{k-1}\phi(x,y) + b\,{}_{k-1}^{\;i}V_2(x,y)\\
{}_k^i V_5(x,y) &= {}_{k-1}^{\;r}V_5(x,y) = c\,{}_{k-1}\phi(x,y) + d\,{}_{k-1}^{\;i}V_5(x,y)
\end{aligned}
\tag{4.32}
$$

These can be substituted back to give a two-step routine but there will still be incident components. These could in turn be replaced to give a three-step routine but again there will be incident terms.

If we impose the condition that $b^2 = d^2$ then all the incident voltage terms cancel and we then obtain an expression in terms of the nodal voltages only:

$$
\begin{aligned}
{}_{k+1}\phi(x,y) = Aa[&{}_k\phi(x-1,y) + {}_k\phi(x,y+1) + {}_k\phi(x+1,y)\\
&+ {}_k\phi(x,y-1)) + Bc\,{}_k\phi(x,y) + (4Aab + Bcd + b^2)\,{}_{k-1}\phi(x,y)
\end{aligned}
\tag{4.33}
$$

We can now take this expression and compare it term-by-term with conventional TLM:

$$
A = \frac{2(R_s + Z_s)}{R + Z + 4(R_s + Z_s)} \tag{4.34}
$$

$$B = \frac{2(R+Z)}{R+Z+4(R_s+Z_s)} \tag{4.35}$$

$$a = \frac{Z}{R+Z} \tag{4.36}$$

$$b = \frac{R-Z}{R+Z} \tag{4.37}$$

$$c = \frac{Z_s}{R_s+Z_s} \tag{4.38}$$

$$d = \frac{R_s-Z_s}{R_s+Z_s} \tag{4.39}$$

The condition $b^2 = d^2$ gives:

$$\frac{RZ}{R_sZ_s} = \frac{R^2+Z^2}{R_s^2+Z_s^2} \tag{4.40}$$

Thus if R, Z and Z_s are chosen then R_s can only take the values ZZ_s/R or RZ_s/Z

The expressions in eqn (4.33) can also be compared term-by-term with finite difference formulations. In the case of comparisons with the diffusion equation we have:

$$Aa = r \quad \text{where } r = \frac{\Delta t}{D\Delta x^2} \tag{4.41a}$$

$$Bc = 1-4r \tag{4.41b}$$

$$4Aab + Bcd + b^2 = 0 \tag{4.41c}$$

The condition $b^2 = d^2$ which is necessary to eliminate incident voltage contributions in the TLM formulation can be satisfied in several ways:

(*i*) $b = d = 0$

gives $R = Z, Rs = Zs$ with $R_s = \dfrac{r}{1-4r}$

When $r=0.25$ then the explicit finite difference in 2-dimensions is operating at its stability limit and $R_s = Z_s = \infty$. The TLM interpretation is that the nodes of the mesh do not have stubs.

(*ii*) $b = -d = 1 - 8r$

implies that $R = Z\left[\frac{1-4r}{4r}\right]$ and $R_s = Z_s\left[\frac{4r}{1-4r}\right]$

Now, if $r=1/8$ then this is identical to the TLM mesh without a stub.

(*iii*) $b = d = -1$

This condition cannot be satisfied by a TLM shunt mesh.

The same approach can be applied to a comparison between the finite difference and TLM formulations for the wave equation and gives very useful information. It confirms that in general if $R = 0$ the two methods provide a solution of the wave equation. With $R = Z$ the two methods provide an identical solution to the diffusion equation. But what of the region $O < R < Z$? We already know that the duFort-Frankel scheme effectively models the telegraphers' equation. This can be expressed in two-dimensions as:

$$
\begin{aligned}
{}_{k+1}\phi(x, y) = &\left[\frac{2r}{1-4r}\right]\{{}_k\phi(x+1, y) + {}_k\phi(x-1, y) + {}_k\phi(x, y+1) + {}_k\phi(x, y-1)\} \\
&+ \left[\frac{1-4r}{1+4r}\right]{}_{k-1}\phi(x, y)
\end{aligned}
\tag{4.42}
$$

A term-by-term comparison with the generalised TLM scheme gives

$$b^2 = d^2 \tag{4.43a}$$

$$Bc = 0 \tag{4.43b}$$

$$4Aab + Bcd + b^2 = \frac{1-4r}{1+4r} \tag{4.43c}$$

$$Aa = \frac{2r}{1+4r} \tag{4.43d}$$

As $Bc=0$ this corresponds to a simple TLM shunt mesh operated without a stub. It is now possible to interpret the inconsistency in the duFort-Frankel method (when used to model the diffusion equation) as being due to the presence of a wave term $(\partial^2\phi/\partial t^2)$.

All of these attempts to develop comparisons between lossy TLM and finite difference formulations of the diffusion equation have so far ignored the question of initial conditions. Unless these are identical in the two approaches then there is no guarantee that they will converge to the same solution.

An initial condition $f(x, y) = {}_0\phi(x, y)$ can be applied to a finite difference approximation of the diffusion equation which can then be stepped forward to give:

$$\begin{aligned}{}_1\phi(x, y) = f(x, y) + r[(f(x+1, y) + f(x-1, y) \\ + f(x, y+1) + f(x, y-1) - 4\, f(x, y)]\end{aligned} \tag{4.44}$$

Stepping forward with the generalised TLM scheme gives:

$$\begin{aligned}{}_1\phi(x, y) = Aa[f(x+1, y) + f(x-1, y) + f(x, y+1) + f(x, y-1)] \\ + Ab[{}^i_kV_3(x-1, y) + {}^i_kV_4(x, y+1) + {}^i_kV_1(x+1, y) \\ + {}^i_kV_2(x, y-1)] + Bc\,f(x, y) + Bd\,{}^i_kV_5(x, y)\end{aligned} \tag{4.45}$$

In order for TLM and finite difference to be equivalent we must have $b^2 = d^2$ and the conditions in eqns (4.41a–4.41c) must be satisfied.

The comparison then forces us to have either $b = d = 0$ or $b = -d = 1-8r$. In the first case this means that it is the nodal values and not the individual pulse values which satisfy the initial conditions. Thus a shunt TLM mesh with no stubs can be excited by pulses on some or all lines with any weighting provided that they sum to $f(x,y)$.

Although the case $b = -d = 1 - 8r$ provides an identical discretisation for both finite difference and TLM, the fact that $b\neq 0$ means that the treatment of the initial conditions would be quite different and therefore the solutions obtained from the two methods would be different.

The eigenvalue solutions of a TLM shunt node

The eigenvalue approach arises from matrix algebra where it is used to find and characterise the non-trivial solutions of a set of linear equations of the form:

$$\begin{bmatrix} a_{11} & a_{12} & a_{13} & \cdot & \cdot \\ a_{21} & a_{22} & \cdot & \cdot & \\ a_{31} & \cdot & \cdot & & \\ \cdot & \cdot & & & \\ \cdot & & & & \end{bmatrix} = \lambda \begin{bmatrix} x_1 \\ x_2 \\ x_3 \\ \cdot \\ \cdot \end{bmatrix} \tag{4.46}$$

This can be re-cast as $(\mathbf{A} - \lambda\mathbf{I})\mathbf{X} = 0$ which has non-trivial solutions so long as the determinant $|\mathbf{A} - \lambda\mathbf{I}| = 0$.

As λ appears once on each row (on the leading diagonal) an n-th order determinant will give rise to a polynomial equation of the form:

$$\lambda^n + c_1\lambda^{n-1} + c_2\lambda^{n-2} + \cdots + c_n = 0 \tag{4.47}$$

This is the characteristic equation and the roots $\lambda_1, \lambda_2, \lambda_3, \ldots, \lambda_n$ are called eigen-values.

This approach has already proved to be particularly useful in lossless TLM. It can be used to determine the modes which propagate on the mesh and also provides information about dispersion. Using such a technique Nielsen and Hoefer [4.17] have demonstrated that as a result of the spatial and temporal sampling process spurious modes may be supported which may corrupt the solution. They identify two types of spurious mode which may be propagated on a 2-d shunt mesh:

type 1 applies to DC(temporal frequency) and all spatial frequencies,
type 2 applies to all spatial frequencies and temporal frequencies given by $k_0\Delta x = \pi$, where $k_0 = 2\pi f_{\text{excitation}}/v_{\text{link}}$ and v_{link} is the velocity of propagation on the link-lines.

They mention that since the network is linear only sources that excite frequency components corresponding to $k_0\Delta x = \pi$ and $k_0\Delta x = 0$ can couple into the spurious modes. As high-frequency excitation is avoided by band-limiting the sources, then only the DC spurious mode is a problem. Sources that shunt the node will not couple to the DC spurious mode, but sources that are placed in series with the link-lines will and should be avoided.

Krumpholz [4.18] is more explicit in identifying the modes in a 2-dimensional shunt node as:

$$\begin{aligned} \lambda_1 &= C_1 + \sqrt{(C_1^2 - 1)} \\ \lambda_2 &= C_1 - \sqrt{C_1^2 - 1)} \\ \lambda_3 &= 1 \\ \lambda_4 &= -1 \end{aligned} \tag{4.48}$$

where $C_1 - \frac{1}{2}[\cos(k_x\Delta x) + \cos(k_y\Delta y)]$

λ_3 and λ_4 represent non-propagating modes and only λ_3 had a physical meaning.

Russer [4.19] has proposed the Alternating TLM (ATLM) scheme which avoids spurious modes in lossless TLM. This is based on work by Sobhy [4.20] which defines the *parity*, p of a network in discretised space (l,m,n) and time (k).

$$p \equiv (k + l + m + n) \bmod 2 \tag{4.49}$$

Since states of even and odd parity are independent of each other, their interconnection may cause the propagation of spurious solutions. If the three-dimensional spatial mesh for conventional TLM is a simple cubic structure of length Δx then the equivalent spatial mesh for ATLM is a face-centred cubic structure with cell length $2\Delta x$. Thus in the absence of stubs we compute only the evolution of states with even parity (sample every $2\Delta t$) and obtain solutions for the odd parity positions by interpolation. This significantly reduces the computational effort while maintaining the accuracy. There are claims that the dispersion properties are improved. Of course in this case dielectric stubs must now be open-circuit lines of length Δx and boundaries must be appropriately located.

Almost simultaneously with this work Pulko *et al* [4.21] had formally observed that observations made at Δt in a link-line lossy mesh of nodal length Δx represent a significant redundancy. There is no loss of accuracy if the same mesh is sampled at intervals of $2\Delta t$. This is not the case if the mesh is expanded to $2\Delta x$ and sampled at $2\Delta t$. The 'error' defined as the difference between the results for the Δx mesh sampled at Δt and $2\Delta t$ display an oscillatory behaviour: "error was zero for ${}_kT(x)_{\Delta t} - {}_kT(x)_{2\Delta t}$ but was significant for ${}_kT(x)_{\Delta t} - {}_{k\pm1}T(x)_{2\Delta t}$." This seems to have close bearing on the 'jumps-to-zero' which have been discussed by several authors [4.10, 4.22, 4.23] and are not necessarily restricted to TLM [4.24]. These are less intrusive for constant voltage or constant current inputs (or their analogues). They are generally absent when a mesh contains stubs or when inhomogeneities lead to inter-nodal scattering.

Although the precise description of the excitation does have a bearing the last observation in the previous paragraph suggests that an eigenvalue comparison between link-line and link resistor configurations might be useful. The scatter and connect processes in a one-dimensional link-line mesh can be coalesced using shift operators as in Soulos *et al* [4.16]:

$$_{k+1}\begin{pmatrix} {}^iV_L \\ {}^iV_R \end{pmatrix} = \begin{pmatrix} \tau\bar{\mathrm{x}} & \rho\bar{\mathrm{x}} \\ \rho\mathrm{x} & \tau\mathrm{x} \end{pmatrix}_k \begin{pmatrix} {}^iV_L \\ {}^iV_R \end{pmatrix} \tag{4.50}$$

x and $\bar{\mathrm{x}}$ are nodal shift operators such that $\mathrm{x}V(i) = V(i+1)$ and $\bar{\mathrm{x}}V(i) = V(i-1)$.

The eigenvalue solutions are $(\tau\mathbf{x} + \tau\bar{\mathbf{x}} \pm \sqrt{\tau^2\mathbf{x} - 2\tau^2\mathbf{x}\bar{\mathbf{x}} + \tau}$ [illegible] $\bar{\mathbf{x}}\mathbf{x})/2$ The equivalent incident matrix for a one-dimensional link [illegible] configuration is:

$$
{}_{k+1}\begin{pmatrix} {}^{i}V_L \\ {}^{i}V_R \end{pmatrix} = \begin{pmatrix} \tau\mathbf{x} & \rho \\ \rho & \tau\bar{\mathbf{x}} \end{pmatrix}_k \begin{pmatrix} {}^{i}V_L \\ {}^{i}V_R \end{pmatrix} \tag{4.51}
$$

and the eigenvalues are: $(\tau\mathbf{x} + \tau\bar{\mathbf{x}} \pm \sqrt{\tau^2\mathbf{x} - 2\tau^2\mathbf{x}\bar{\mathbf{x}} + \tau^2\bar{\mathbf{x}}\bar{\mathbf{x}} + 4\rho^2})/2$

On the basis that $\mathbf{x}\bar{\mathbf{x}} = \bar{\mathbf{x}}\mathbf{x}$ it might be argued that there is no difference in these two results. This is not correct because we are not comparing like with like. If we take the space and time discretisations in a link-line configuration as datum points then we can draw equivalences for the link-resistor node and these are shown in figure 4.6.

Using figure 4.6 it can be seen that the nodal voltages in a link resistor model are computed at the half-time intervals. Thus using the link-line model as the reference point we can write:

$$
{}_{k+1/2}\phi(x') = {}_{k+1/2}^{\;\;i}V_L(x') + {}_{k+1/2}^{\;\;i}V_R(x') \tag{4.52}
$$

Immediately afterwards ${}_{k+1/2}^{\;\;i}V_L(x')$ becomes ${}_{k+1/2}^{\;\;r}V_R(x')$ and ${}_{k+1/2}^{\;\;i}V_R(x')$ becomes ${}_{k+1/2}^{\;\;r}V_L(x')$.

The reflection and transmission at the resistors provide the incoming pulses for $k + 3/2$.

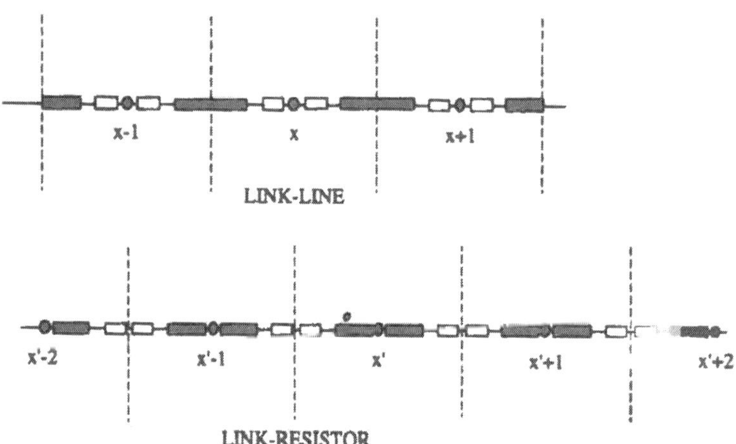

Figure 4.6 The spatial relationships between link-line and link-resistor nodal discretisations of the same sample. It can be seen that $x' = x + \Delta x/2$.

We can try to equate like with like:

$$ {}_{k+1/2}^{\;s}V_L(x') = {}_k^iV_R(x) \text{ in the link-line model} $$
$$ {}_{k+1/2}^{\;s}V_R(x') = {}_k^iV_L(x+1) $$

or shifting: ${}_{k+1/2}^{\;s}V_R(x'-1) = k^{\,i}V_L(x)$

$$ {}_{k+1/2}^{\;i}VR(x') = {}_k^sV_L(x+1) = \rho\,{}_k^iV_L(x+1) + \tau\,{}_k^iV_R(x+1) $$
$$ {}_{k+1/2}^{\;i}V_L(x') = {}_k^sV_R(x) = \rho\,{}_k^iV_R(x) + \tau\,{}_k^iV_L(x) $$

or

$$ {}_{k+1/2}\begin{pmatrix} {}^iV_L(x') \\ {}^iV_R(x') \end{pmatrix} = \begin{pmatrix} \tau & \rho \\ \rho\mathrm{x} & \tau\mathrm{x} \end{pmatrix}_k \begin{pmatrix} {}^iV_L(x) \\ {}^iV_R(x) \end{pmatrix} \tag{4.53} $$

The eigenvalues for the scatter-connect matrix are then:

$$ \frac{\tau}{2} + \frac{\tau\mathrm{x}}{2} \pm \sqrt{\tau^2 - 2\tau^2\mathrm{x} + \tau^2\mathrm{xx} + 4\rho^2\mathrm{x}} \tag{4.54} $$

This is as if the link-resistor formulation consisted of a mesh of length $\Delta x/2$ which was only sampled at every second time-step (where time-step $= \Delta t/2$) at 'even' locations. In other words, the link-resistor formulation is equivalent to an ATLM or non-redundant scheme for a mesh of length $\Delta x/2$.

A state-space interpretation of diffusive TLM

'Discrete state-space' is a system concept which has its origins in control theory [4.25, 4.26]. The state of the system is defined as a set of variables such that a knowledge of these at some time together with a knowledge of future inputs is sufficient to provide a complete description of the behaviour of the system.

If the system state is described by an n-dimensional vector, $x(t)$ and if there are input and output vectors $u(t)$ and $y(t)$ then:

$$ x(t+1) = f(x(t), u(t), t) \quad \text{and} \quad y(t) = g(x(t), u(t), t) $$

The description of the system is then a matter of identifying an appropriate vector of state variables, x and determining the functions f and g corresponding to the dynamic behaviour of the system.

In a system where the super-position principle holds we can write:

$$f(x(t), u(t), t) = Ax(t) + Bu(t) = x(t+1) \tag{4.55}$$
$$g(x(t), u(t), t) = Cx(t) + Du(t) = y(t) \tag{4.56}$$

Witwit *et al* [4.27] have applied this concept to TLM and using a range of examples they have demonstrated that it is possible to draw a one-to-one mapping between conventional algorithms and state-space descriptions. We will now use a variant of their approach to develop discrete state-space expressions for a TLM link-line diffusion model.

The TLM scatter-connect processes in presence of excitation can be expressed as:

$$\begin{bmatrix} \mathbf{Tx} & 0 \\ 0 & \mathbf{T\bar{x}} \end{bmatrix} \begin{bmatrix} {}_k^iV_1(x) \\ {}_k^iV_2(x) \end{bmatrix} = \begin{bmatrix} \tau & \rho \\ \rho & \tau \end{bmatrix} \begin{bmatrix} {}_k^iV_1(x) \\ {}_k^iV_2(x) \end{bmatrix} + \begin{bmatrix} b_1 \\ b_2 \end{bmatrix} {}_ku(x) \tag{4.57}$$

In this case $\mathbf{T}$ is the time-shift operator, $\mathbf{x}$ and $\bar{\mathbf{x}}$ are the spatial shift operators (as described previously), L and R from previous sections have been replaced by generalised directions, 1 and 2. $b1$ and $b2$ are the coefficients of the B matrix in eqn (4.55) and $u(x)$ is the nodal input at x.

The nodal output is then

$${}_ky(x) = [c1, c2] \begin{bmatrix} {}_k^iV_1(x) \\ {}_k^iV_2(x) \end{bmatrix} \tag{4.58}$$

where $c1$ and $c2$ are the coefficients of the output matrix C in eqn (4.56) The nodal output can be related to the nodal input via a transfer function, G

$${}_ky(x) = G\,{}_ku(x) \tag{4.59}$$

which is obtained by eliminating the state vector i.e. the vector of incident pulses from eqns (4.57) and (4.58).

Eqn (4.57) can be rearranged to give:

$$\begin{bmatrix} \mathbf{Tx} & 0 \\ 0 & \mathbf{T\bar{x}} \end{bmatrix} \begin{bmatrix} {}_k^iV_1(x) \\ {}_k^iV_2(x) \end{bmatrix} - \begin{bmatrix} \tau & \rho \\ \rho & \tau \end{bmatrix} \begin{bmatrix} {}_k^iV_1(x) \\ {}_k^iV_2(x) \end{bmatrix} = \begin{bmatrix} b_1 \\ b_2 \end{bmatrix} {}_ku(x) \tag{4.60}$$

$\begin{bmatrix} {}_k^iV_1(x) \\ {}_k^iV_2(x) \end{bmatrix}$ can then be removed from eqn (4.57) to give:

$${}_ky(x) = [c_1, c_2] \left[\begin{bmatrix} \mathbf{Tx} & 0 \\ 0 & \mathbf{T\bar{x}} \end{bmatrix} - \begin{bmatrix} \tau & \rho \\ \rho & \tau \end{bmatrix} \right]^{-1} \begin{bmatrix} b_1 \\ b_2 \end{bmatrix} {}_ku(x) \tag{4.61}$$

so that

$$G = [c_1, c_2]\left[\begin{bmatrix} \mathbf{Tx} & 0 \\ 0 & \mathbf{T\bar{x}} \end{bmatrix} - \begin{bmatrix} \tau & \rho \\ \rho & \tau \end{bmatrix}\right]^{-1} \begin{bmatrix} b_1 \\ b_2 \end{bmatrix} \tag{4.62}$$

This can be multiplied out and when the denominator is set equal to zero we obtain the characteristic equation of the process being modelled. With this done we obtain:

$$(\mathbf{T} - 2 + \mathbf{T}^{-1}) + 2\rho(1 - \mathbf{T}^{-1}) = \tau(\mathbf{x} - 2 + \bar{\mathbf{x}}) \tag{4.63}$$

We can now observe some discrete equivalences

$$\frac{(\mathbf{T} - 2 + \mathbf{T}^{-1})}{\Delta t^2} \equiv \frac{\partial^2}{\partial t^2} \tag{4.64a}$$

$$\frac{(1 - \mathbf{T}^{-1})}{\Delta t} \equiv \frac{\partial}{\partial t} \tag{4.64b}$$

$$\frac{(\mathbf{x} - 2 + \bar{\mathbf{x}})}{\Delta x^2} \equiv \frac{\partial^2}{\partial x^2} \tag{4.64c}$$

If these are substituted back into eqn (4.63) we obtain

$$\frac{\partial^2}{\partial t^2} + \left[\frac{2\rho}{\Delta t}\right]\frac{\partial}{\partial t} = \tau\left[\frac{\Delta x^2}{\Delta t^2}\right]\frac{\partial^2}{\partial x^2} \tag{4.65}$$

a damped wave equation.

If on the other hand we had set $\rho = \tau = R/(R + Z)$ in eqn (4.63) it would then reduce to:

$$(\mathbf{T} - 1) = \frac{(\mathbf{x} - 2 + \bar{\mathbf{x}})}{2} \tag{4.66}$$

using our equivalences we have;

$$\Delta t\frac{\partial}{\partial t} = \left[\frac{\Delta x^2}{2}\right]\frac{\partial^2}{\partial x^2} \text{ or } \frac{\partial}{\partial t} = \left[\frac{\Delta x^2}{2\Delta t}\right]\frac{\partial^2}{\partial x^2} \tag{4.67}$$

Thus lossy TLM is totally amenable to a discrete state-space description and for general scattering the characteristic equation of the state-space description yields a damped wave equation. When $R = Z$ then the result is an exact correspondence with the diffusion equation.

Lossy TLM as a probability process

One of the major problems which has faced those researching the physical meaning of finite difference schemes has been the *infinite velocity paradox.* A rigorous analysis suggests that at any time $t > 0$ there will be a potential due the initial excitation at all points throughout the network. If this unfortunate situation is ignored then it is possible to treat finite differences in terms of probabilities. Particularly when we are at the stability limit for explicit finite difference then the problem becomes equivalent to a simple random walk. Of course there are two ways of looking at this:

- we could consider an individual particle which had an equal probability of moving to left or right in a one-dimensional formulation. We could chart its history over time and we could repeat the experiment many times.
- we could consider an ensemble of many particles each of which could move independently at every time-step. In this case we could say that approximately half of the population moved to left and half to right at each time-step.

We will be working with the latter scheme and we will also be considering situations where we have biased random walks. We will indicate how the infinite velocity paradox does not arise if random walks are interpreted as a process on a TLM mesh. Counterwise, if TLM can be treated as a random walk process, then the TLM scattering parameters can be seen as probabilities. This then allows us to devise closed algebraic forms of what is normally an iterative process. The work by de Cogan and Enders [4.10, 4.24, 4.28] drew its inspiration from the much neglected work of Taitel [4.29] and of Goldstein [4.30], whose TLM-like solution of the telegraphers' equation predates TLM by 20 years.

Simple random walks with finite time-step

The Taitel treatment considers a sample divided into N nodes of equal length Δx with time discretised into units Δt, being the time of flight between two adjacent nodes.

$$_{k+1}T(x) = \frac{_{k}T(x-1) + {_{k}T(x+1)}}{2} \tag{4.68}$$

If both sides are presented in terms of Taylor expansions then we obtain

$$\frac{\partial}{\partial t}T(x,t)+\frac{\Delta t}{2}\frac{\partial^2}{\partial t^2}T(x,t)=\frac{\Delta x^2}{2\Delta t}\frac{\partial^2}{\partial x^2}T(x,t)+\ldots \tag{4.69}$$

Within the first and second order this comprises the parabolic and hyperbolic heat equations respectively, where the flux relaxation time, τ_r is set at $\Delta t/2$. By construction, the propagation increases with the flux speed and the temperature always retains its maximal value at the sample boundary, if that was the condition at t = 0.

The factor $\Delta x^2/2\Delta t$ is equivalent to the diffusion constant and can therefore be described as $k_T/(\rho\ C_p)$ where k_T = thermal conductivity, ρ = density, and C_p = specific heat. Thus eqn (4.68) is a special case of the standard 1-step explicit finite-difference approximation to the linear homogeneous diffusion equation.

It is also possible to deduce eqn (4.68) in terms of heat flux using the discrete constitutive law:

$$J\left(x-\frac{1}{2},k+\frac{1}{2}\right)=-\frac{k_T}{\Delta x}\left[{}_kT(x)-{}_kT(x-1)\right] \tag{4.70}$$

and the discrete continuity equation:

$$\frac{\rho C_p}{\Delta t}\left[{}_kT(x)-{}_kT(x)\right]+\frac{1}{\Delta x}\left[J\left(x+\frac{1}{2},k+\frac{1}{2}\right)-J\left(x-\frac{1}{2},k+\frac{1}{2}\right)\right]=0 \tag{4.71}$$

In this context it should be remembered that eqn (4.70) is the discrete analogue of the constitutive law of hyperbolic heat diffusion theory.

$$\tau_r\frac{\partial}{\partial t}J(x,t)+J(x,t)=-k_T\frac{\partial}{\partial x}T(x,t) \tag{4.72}$$

where $J(x-\frac{1}{2},k+\frac{1}{2})=\frac{1}{2}\left[J(x-\frac{1}{2},k)+J(x-\frac{1}{2},k+1)\right]$.

In other words, while temperatures are registered at discrete points in space and time, heat 'current' is monitored at the half intervals.

Mesh excitations in a one-dimensional simple random walk model

Using the discrete diffusion equation (4.68) as a starting point it is possible to develop closed algebraic solutions for some well-known standard excitations:

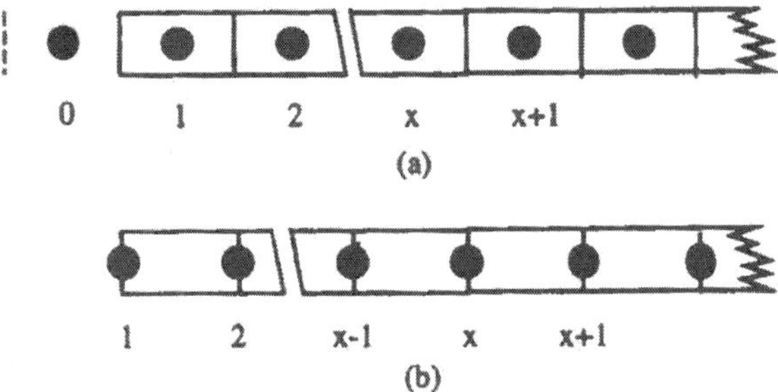

Figure 4.7(a) Material boundaries located at cell boundaries with nodes at the centre of each cell, (b) material boundaries located at the nodes.

(1) Flash-light excitation

Dirac-like excitation can be described by a combination of Kronecker initial data and adiabatic boundary conditions; the latter expressing vanishing surface flux.

$$
\begin{aligned}
{}_0T(x) &= \delta(x,1) \quad (x = 1,2,\ldots) \\
{}_{k+1}T(1) &= \frac{{}_kT(1) + {}_kT(2)}{2}
\end{aligned}
$$

The general solution to this is

$$
{}^{f}_{k}T(x) = \frac{1}{2^k}\binom{k}{\left[\frac{k-x+1}{2}\right]} \tag{4.73}
$$

The square bracket term represents the integer value and the outer (round) parentheses imply a binomial coefficient $\binom{a}{b} = a!/b!(a-b)!$
By reference to figure 4.7(a) this can be easily proved by showing that eqn (4.73) satisfies eqn (4.68), satisfies the boundary conditions and guarantees spatial heat conservation.

(2) Constant surface temperature excitation

In this treatment the temperature values are normalised to the surface temperature T_s by setting $T(0,k) = 1$ for all time. The resulting temperature ${}^{T}T(x,k)$ is the sum of the effects of a series of flash excitations (running index, j) which have reached position x at time k. This derivation uses the mesh in figure 4.7(b).

$$ {}^{T}{}_{k}T(x) = \frac{1}{2^k}\sum_{j=0}^{k-x}\binom{k}{\left[\frac{k-x-j}{2}\right]} = \frac{1}{2^k}\sum_{j=0}^{k-x}\binom{k}{\left[\frac{j}{2}\right]} \tag{4.74} $$

The proof is analogous to the previous case.

(3) Constant surface flux

In this case the temperature may be normalised to $(\Delta x \,.\, Js/kT)$, where J_s is the surface flux density. In terms of temperature one can write ${}_0T(0) = 1$ and ${}_0T(x) = 0$ $(x = 1, 2, 3, ...)$ with the condition ${}_kT(0) = {}_kT(1) + 1$ at the surface. In contrast to the previous two cases there is no simple law of construction which allows one to write a generalised form by inspection of the first results. However, it should be realised that the temperature at the node nearest the surface can be expressed using eqn (4.68) as:

$$ {}_{k+1}T(1) = \frac{{}_kT(0) + {}_kT(2)}{2} = \frac{1 + {}_kT(1) + {}_kT(2)}{2} \tag{4.75} $$

The first of these terms represents a discretely continuous Kronecker excitation at node 1 and the others represent the adiabatic boundary condit he light of this one can write:

$$ {}^{J}{}_{k}T(x) = \frac{1}{2}\sum_{j=1}^{k}{}^{fl}{}_{k-j}T(x) = \sum_{j=0}^{k-x}\frac{1}{2^{x+j}}\binom{j+1-x}{\left[\frac{j}{2}\right]} \tag{4.76} $$

The proof is as in the previous cases except that heat conservation is expressed by the relation:

$$ \sum_{x=1}^{\infty}{}^{J}{}_{k}T(x) = \frac{k}{2} \tag{4.77} $$

The factor 1/2 arises by virtue of the fact that the distance from node 1 to the surface is $\Delta x/2$.

Propagation processes in a two-dimensional simple random walk model

If the point of flash-light excitation at $t = 0$ is at $x = 0$, $y = 0$ then visual inspection of the propagation process on a two dimensional mesh shows that it follows what is called the *Bernoulli trial*:

(1) the population at position (x,y) at time k is zero for $|x| + |y| > k$
(2) the population at position (x,y) at time k is zero for $k - x - y$ odd.
(3) For $|x| + |y| \leq k$ <u>and</u> $(k - x - y)$ even the population is given by:

$$\frac{(\text{initial input})}{4^k}\binom{k}{\left[\frac{k-x+y}{2}\right]}\binom{k}{\left[\frac{k-x-y}{2}\right]} \tag{4.78}$$

It can be seen that the jumps-to-zero are observed it the time interval between injection and inspection for a given location is odd. However, if there is injection at two times which are an odd interval apart then there will be no jumps-to-zero. We could check this over many values of k by looking at the value at $x = 0$, $y = 0$ for injections of magnitude 1024 input at $x = 0$, $y = 0$ at $k = 0$ and $k = 1$.

These concepts could be extended to include a situation where a numerical value (1024 for convenience) representing a concentration of material is injected at time $k=0$ and also at time $k=1$. Conditions (2) and (3) above tell us that it would be:

$$\frac{1024}{4^k}\binom{k}{\left[\frac{k}{2}\right]}\binom{k}{\left[\frac{k}{2}\right]} \quad k = 0, 2, 4, 6, \ldots \text{ and}$$

$$\frac{1024}{4^{k-1}}\binom{k-1}{\left[\frac{k-1}{2}\right]}\binom{k-1}{\left[\frac{k-1}{2}\right]} \quad k = 1, 3, 5, 7, \ldots$$

We can now develop an expression for constant flux excitation. For example, a heat source which leads to a temperature rise at every time-step can be treated as an ensemble of independent flash-light excitations whose contributions are summed.

The simple expression can be extended to cover the response at location (x,y) and time k due to a flash-light excitation at location (x',y') and time k. so long as $|x - x'| + |y - y'| \leq (k - k')$ and $(k - k') - (x - x') - (y - y')$ is even then:

$$\begin{aligned}{}_kT(x, y) = \frac{1}{4^{(k-k')}} &\binom{k-k'}{\left[\frac{(k-k')-(x-x')+(y-y')}{2}\right]} \times \\ &\binom{k-k'}{\left[\frac{(k-k')-(x-x')-(y-y')}{2}\right]} T_0(x', y', k')\end{aligned} \tag{4.79}$$

Otherwise it is zero. $T_0(x',y',k')$ is the initial excitation.

The concept has very close similarities to Green's functions in continuous mathematics and we therefore define a ***discrete Green's function*** as:

$$
\begin{aligned}
&{}_{k-k'}G(x, y;\, x', y') \\
&= \frac{1}{4^{(k-k')}}\left(\begin{matrix} k-k' \\ \left[\dfrac{(k-k')-(x-x')+(y-y')}{2}\right]\end{matrix}\right)\left(\begin{matrix} k-k' \\ \left[\dfrac{(k-k')-(x-x')-(y-y')}{2}\right]\end{matrix}\right)
\end{aligned}
\tag{4.80}
$$

Thus, in a problem involving multiple inputs in space and time each input is regarded as propagating independently and the resultant temperature at any position and time is the sum of the contributions from the individual discrete Green's functions weighted by the corresponding input strengths, ${}_{k'}T^{s}(x',y')$.

$$
{}_{k}T(x, y) = \sum_{k'=0}^{k}\ \sum_{y'=-\infty}^{+\infty}\ \sum_{x'=-\infty}^{+\infty} {}_{k-k'}G(x, y;\, x', y')\,{}_{k'}T^{s}(x', y') \tag{4.81}
$$

Simple random walks in inhomogeneous problems

In the analysis of inhomogeneous materials of Enders and de Cogan [4.31] it was assumed that there was no thermal resistance between materials. For short-hand we will define the heat capacity per unit volume, $p(x) = \rho(x)\, C_p(x)$ and we will also need to define another parameter, the effusivity. Eqn (4.70) can be rewritten as

$$
J\left(x - \frac{1}{2},\ k + \frac{1}{2}\right) = -\frac{e}{\sqrt{2\Delta t}}\left[T(x, k) - T(x-1, k)\right] \tag{4.82}
$$

where $e = \sqrt{\rho\, C_p k_T}$ is the effusivity or heat penetration coefficient. Like the thermal diffusivity, $k_T/\rho C_p$, it expresses the fact that thermal diffusion is an interplay of heat storage and heat transfer. Two simple cases can now be considered:

(1) Layer interface situated between nodes

This situation corresponds to the mesh configuration shown in figure 4.7(a). The heat conservation law for adiabatic boundary conditions in a system of layered material with nodes of different thermal properties requires that:

$$\sum_{x} p_x \Delta x_x \, T(x,k) = \text{constant} \tag{4.83}$$

where the mesh cell length of node x is given by $\Delta x_x = \sqrt{2\Delta t\, D_x}$ (D_x is the diffusion constant of the material of node x). This equation can also be expressed in terms of effusivity as:

$$\sqrt{2\Delta t}\sum_{x} e_x T(x,k) = \text{constant} \tag{4.84}$$

Using eqn (4.68) this condition is satisfied by:

$$\begin{aligned}2T(x,k) = (1+r_{x-})T(x-1,k) + (1-r_{x+})T(x+1,k)\\ +(r_{x+}-r_{x-})T(x,k)\end{aligned} \tag{4.85}$$

where the material interface reflection coefficients r_{x-} and r_{x+} are defined as:

$$r_{x-} = \frac{e_{x-1}-e_x}{e_{x-1}+e_x} \text{ and } r_{x+} = \frac{e_x - e_{x+1}}{e_x + e_{x+1}}$$

As mentioned above, materials with equal effusivity behave like a single homogeneous bulk. It should be noted that since the nodes are located at the centre of homogeneous cells, the transition probability of going to the left or to the right is still one-half. It is really more precise to think of the quasi particles as carriers of a hydrodynamic current density, pTv_0 ($v_0 = \Delta x/\Delta t$), since it is in fact this flux which travels in equal part to left and right.

The heat flux between disparate nodes $x-1$ and x is calculated by equating that from node $x-1$ to the interface with that from the interface to node x:

$$\frac{k_T(x-1)}{\Delta x/2}[T_i - T(x-1,k) = \frac{k_T(x)}{\Delta x/2}[T(x,k) - T_i] \tag{4.87}$$

This allows us to define a value for the average interface temperature for the boundary:

$$T_i(k) = \frac{e_{x-1}T(x-1,k) - e_x T(x,k)}{e_{x-1}+e_x} \tag{4.88}$$

And from this one gets:

$$J\left(x-\frac{1}{2},k+\frac{1}{2}\right) = \sqrt{\frac{2}{\Delta t}}\left[\frac{e_{x-1}e_x}{e_{x-1}+e_x}\right][T(x,k) - T(x-1,k)] \tag{4.89}$$

(2) Layer interface situated at a node

When the spatial discretisation is arranged so that there is a node located at the interface between two materials the generalised laws of motion (eqn (4.68)) then becomes

$$T(x,k+1)=\frac{e_{x-}T(x-1,k)+e_{x+}T(x+1,k)}{e_{x-}+e_{x+}} \tag{4.90}$$

In this case e_{x-} refers to the effusivity of the material to the left $(\Delta x/2)$ interval of the node and e_{x+} refers to the effusivity of the material to the right $(\Delta x/2)$ interval of the node. With this in mind we can define reflection a coefficient specific to node x as $r(x)=(e_{x-}-e_{x+})/(e_{x-}+e_{x+})$ so that eqn (4.90) can be restated as

$$T(x,k+1)=\frac{(1+r(x))T(x-1,k)+(1-r(x+1))T(x+1,k)}{2} \tag{4.91}$$

The heat-flux is given by

$$J\left(x-\frac{1}{2},k+\frac{1}{2}\right)=\frac{-e_{x-}}{\sqrt{2\Delta t}}(T(x,k)-T(x-1,k)) \tag{4.92}$$

Eqn. (4.90) obeys the heat conservation law

$$\frac{1}{2}\sum_x(p_{x-}\Delta x_{x-}+p_{x+}\Delta x_{x+})T(x,k)=\sqrt{\frac{\tau}{2}}\sum_x(e_{x-}+e_{x+})T(x,k) \tag{4.93}$$

which is constant. In contrast to eqn (4.87) the flux in the eqn (4.92) contains no reflection, since the material between two nodes is assumed to be homogeneous.

Biased or correlated random walks with finite time-step

The Taitel treatment considers a simple random walk model. For many purposes this is too simple, in particular, the restriction in the time-step limits its range of application. Goldstein [4.30] has proposed an approach which assumes that the probability of a diffusing particle at position x at time k is the sum of the probabilities of crossing cell boundaries from left, $L(x,k)$ and from right, $R(x,k)$. In thermal terms the temperature is then given by

$$T(x,k) = L(x,k) + R(x,k) \tag{4.94}$$

The formulation implicitly includes the possibility of material inhomogeneity and once again two cases can be considered.

(1) Scattering at the node

Within the context of heat-flow in a one-dimensional layered material the configuration with nodes positioned at the boundaries will give rise to a scattering equation of the form

$$\begin{aligned} L(x,k+1) &= \rho_L(x)R(x-1,k) + \tau_L(x)L(x-1,k) \\ R(x,k+1) &= \rho_R(x)L(x+1,k) + \tau_R(x)R(x+1,k) \end{aligned} \tag{4.95}$$

where ρ and τ are the reflection and transmission coefficients at the node, x. A quick inspection will show that this gives rise to the jumps-to-zero. These can be avoided by having the scattering boundaries at the mid-points between nodes.

(2) Scattering between nodes

In this case the scattering equations will give

$$\begin{aligned} L(x,k+1) &= \rho_L(x)R(x,k) + \tau_L(x)L(x-1,k) \\ R(x,k+1) &= \rho_R(x)L(x,k) + \tau_L(x+1)R(x+1,k) \end{aligned} \tag{4.96}$$

where ρ_L, ρ_R, τ_L, and τ_R are the reflection and transmission coefficients to the left and right hand side of each node. For node x these are defined as

$$\begin{aligned} \rho_L(x) &= \frac{\Delta x/k_T(x) + \Delta x/k_T(x-1) + \Delta t/(\Delta x\ p(x-1)) - \Delta t/(\Delta x\ p(x))}{\Delta x/k_T(x) + \Delta x/k_T(x-1) + \Delta t/(\Delta x\ p(x-1)) - \Delta t/(\Delta x\ p(x))} \\ &= 1 - \tau_R(x-1) \\ \rho_L(x) &= \frac{\Delta x/k_T(x) + \Delta x/k_T(x+1) + \Delta t/(\Delta x\ p(x+1)) - \Delta t/(\Delta x\ p(x))}{\Delta x/k_T(x) + \Delta x/k_T(x+1) + \Delta t/(\Delta x\ p(x+1)) - \Delta t/(\Delta x\ p(x))} \\ &= 1 - \tau_L(x+1) \end{aligned}$$

Note that $\rho_L(x)$ and $\tau_L(x)$ do not sum to unity; nor do $\rho_R(x)$ and $\tau_R(x)$. However heat conservation forces $\tau_R(x-1)p(x) = \tau_R(x)p(x-1)$.

The heat flux density from cell $(x-1)$ to (x) during the time interval $k\Delta t$ and $(k+1)\Delta t$ can be determined using eqns (4.96) and the constitutive equation of hyperbolic heat transfer theory,

$-k_T(x)\partial T(x,t)/\partial x = J(x,t)+\tau_r(x)\ \partial J(x,t)/\partial t$ where τ_r is the heat flux relaxation time, which represents the inertia of matter with respect to perturbation and this leads to a finite speed of propagation of perturbations. With $u_0=\Delta x/\Delta t$ this yields

$$\begin{aligned} J\left(x-\frac{\Delta x}{2},k+\frac{\Delta t}{2}\right) &= J\left((x-1)+\frac{\Delta x}{2},k+\frac{\Delta t}{2}\right) \\ &= u_0 p(x)\tau_R(x-1)[L(x-1,k)-R(x,k)] \end{aligned} \tag{4.97}$$

It turns out that Δx and Δt are no longer connected through the diffusivity.

Eqns (4.96) are separable and can be presented as

$$\begin{aligned} R(x,k+1) &= \tau_L(x+1)R(x+1,k)+\frac{\rho_R(x)}{\rho_R(x-1)}\tau_R(x-1)R(x-1,k) \\ &\quad +\frac{\rho_R(x)}{\rho_R(x-1)}[\rho_R(x-1)\rho_L(x)-\tau_R(x-1)\tau_L(x)]R(x,k-1) \\ L(x,k+1) &= \tau_R(x-1)L(x-1,k)+\frac{\rho_L(x)}{\rho_L(x+1)}\tau_L(x+1)L(x+1,k) \\ &\quad +\frac{\rho_L(x)}{\rho_L(x+1)}[\rho_L(x+1)\rho_R(x)-\tau_L(x+1)\tau_L(x)]L(x,k-1) \end{aligned} \tag{4.98}$$

The third term in both of these equations indicates that there is a correlation between successive time-steps and hence the title *Correlated Random Walk*. This is a correlation between the motion during two subsequent time steps – the model is two-step Markovian and to that extent can be analysed using Markov techniques. In general eqns (4.98) cannot be condensed into a single equation for $T(x,k)$, as eqn (4.94). Thus, in practise, we work with the system (4.95) and (4.96) respectively.

If we turn to consider a uniform, homogeneous material then $\rho_L(x)=\rho_R(x)=\rho, \tau_L(x)=\tau_R(x)=\tau$ and if we resume the TLM-like formulations then the combination of eqns (4.94) and (4.95) gives:

$${}_{k+2}V(x)=\tau[{}_{k+1}V(x-1)+{}_{k+1}V(x+1)]+(\rho^2-\tau^2)\,{}_kV(x) \tag{4.99}$$

which clearly indicates the two-step correlated motion

We can now proceed to undertake a one-to-one mapping between the probabilistic parameters which have been outlined above and those of lossy TLM. L_d, C_d and R_d are the distributed parameters of inductance, capacitance and resistance in the electrical network analogue (not to be confused with $R(x,k)$ and $L(x,k)$ of the previous section). Inductance and capacitance are subsumed in the impedance $Z = \sqrt{L_d/C_d}$ and the velocity of propagation of signals on the network is given by $v_0 = \sqrt{1/L_dC_d}$. The scattering parameters are:

$$\rho_L(x)_{\text{TLM}} = \frac{Z(x-1)+R(x-1)+R(x)-Z(x)}{Z(x-1)+R(x-1)+R(x)+Z(x)}$$
$$\rho_R(x)_{\text{TLM}} = \frac{Z(x+1)+R(x+1)+R(x)-Z(x)}{Z(x+1)+R(x+1)+R(x)+Z(x)} \tag{4.100}$$

It is immediately possible to draw links between the TLM parameters and the thermal parameters which have been used in this presentation:

$$R(x)_{\text{TLM}} = \frac{\Delta x}{A\,k_T(x)} \qquad C(x)_{\text{TLM}} = \Delta x\,A\,p(x)$$
$$L(x)_{\text{TLM}} = \frac{\Delta x\,\tau_r(x)}{A\,k_T(x)} \qquad Z(x) = \sqrt{\frac{L(x)}{C(x)}} = \frac{\sqrt{\tau_r(x)}}{A\,e(x)} \tag{4.101}$$

where A is the cross-sectional area.

Of particular importance is the relationship between the inductance, and the flux relaxation time which provides a physical interpretation of L within the context of diffusion. This is not surprising as they both represent inertia properties. This is also consistent with standard concepts of electrical theory. The inductance and resistance expressions in eqn (4.101) can be amalgamated to give:

$$\tau_r(x) = \frac{L(x)_{\text{TLM}}}{R(x)_{\text{TLM}}} \tag{4.102}$$

which is analogous to the time constant in an LR electrical circuit where the rise and decay of current are respectively:

$$I(t) = I_0\left[e^{-\frac{t}{\tau}}+1\right] \quad \text{and} \quad I(t) = I_0 e^{-\frac{t}{\tau}} \tag{4.103}$$

Consequently, with the addition of properly defined transition (reflection, transmission) coefficients at boundaries, this treatment constitutes at once the correlated random walk model of Brownian motion and the TLM numerical routine for computing one dimensional scalar propagation. There is also a suggestion that whereas R and C are the proper parameters when TLM algorithms are couched in terms of temperature, L and R may be the more appropriate parameters if an algorithm is being used to derive heat flux.

Closed solutions of the TLM scattering equations

Now that we have demonstrated that there are isomorphisms between TLM, correlated random walks and Green's function approaches we will use these to devise closed solutions for the TLM scattering processes in one-dimensional homogeneous material.

Eqn (4.94) indicates that the sum of the left- and right-going pulses (eqn(4.95)) constitutes the nodal voltage.

The *fundamental TLM* solution, ${}_kV_{(x)}^{TLM}$, is called that solution of these equations which satisfies the initial conditions

$$
\begin{aligned}
{}_0V(x) &= \delta(x,0) \\
{}_1V(x) &= \frac{\rho+\tau}{2}(\delta(x,1)+\delta(x,-1)); \quad i=-\infty\ldots+\infty
\end{aligned} \tag{4.104a}
$$

$$
{}_0^iV_L(x) = {}_0^iV_R(x) = \frac{1}{2}\,\delta(x,0); \quad i=-\infty\ldots+\infty \tag{4.104b}
$$

and the boundary conditions

$$
{}_kV(-\infty) = {}_kV(\infty) = 0; \quad k=0,1,2,\ldots \tag{4.105}
$$

Elimination of the left and right incident pulses from the scattering equations yields a single (decoupled) two-step equation which is identical to eqn (4.99) and is equivalent to the duFort-Frankel scheme (eqn ((4.11)).

The development of closed-form solutions can be easily approached if, for the time-being, one discards the initial conditions (eqn (4.104a)) for time $k=1$. Then eqn (4.99) can be converted into the equivalent integral equation

$$
{}_{k+2}V(i) = {}_{k+2}V^{(0)}(i) + (\rho^2-\tau^2)\sum_{k'=0}^{k}\sum_{i'=-k'}^{k'} {}_{k-k'}V^{(0)}(i-i')\;{}_{k'}V(i') \tag{4.106}
$$

where

$$
{}_kV^{(0)}(i) = 0 \text{ for } |i|>k \text{ or } (k-i) \text{ odd} \tag{4.107a}
$$

$$
{}_kV^{(0)}(i) = \tau^k \binom{k}{\left[\frac{k-i}{2}\right]} \text{ for } k-|i|\geq 0 \text{ and even} \tag{4.107b}
$$

We can see that much of this is closely related to the Green's function approach in simple random walk models.

Eqn (4.106) is called the *Dyson equation* and is used to obtain a result (called *the fundamental solution*, V^f) for the situation with the altered boundary condition.

This is obtained by means of successive iteration as:

$$ {}_kV^f(x) = \sum_{m=0}^{n} {}_kV^{(m)}(x) \quad n \equiv \frac{k-|i|}{2} \text{ (entire, non-negative)} \tag{4.108a}$$

$$ {}_kV^{(m)} = (\rho^2 - \tau^2)^m \binom{k-m}{m} {}_{k-2m}V^{(m)}(x) \quad m = 0, 1, \ldots, n \tag{4.108b}$$

Thus

$$ {}_kV^f(x) = \tau^k \binom{k}{n} {}_2F_1(-n, -n-i; -k; 1-r^{-2}) \quad \text{where } r \equiv \tau/\rho \tag{4.109a}$$

or

$$ {}_kV^f(x) = \tau^k {}_2F_1(-n, -n-i; 1; r^{-2}) \tag{4.109b}$$

where ${}_2F_1$ is a Gauss hypergeometric function which can be rewritten as a Jacobi polynomial for the given argument values.

These are closed polynomial expressions for the fundamental solution of eqn (4.99) under the initial boundary condition (4.104a) for $k = 0$, but it will be remembered that this was discarded for $k = 1$ when we had in fact

$$ {}_1V^f(x) = {}_1V^{(0)}(x) = \tau(\delta(x,1) + \delta(x,-1)) \tag{4.110}$$

In eqn (4.109a) the effect of correlation (*i.e.* $r \neq 1$) is expressed as a fraction of $V^{(0)}$ as it is given in eqn (4.107) and is just a single hypergeometric function. The alternative, (4.109b) is a convenient form for performing the continuum limit.

In passing it is noted that, by a conceptually similar argument, we could rewrite eqn (4.99) as:

$$ {}_{k+2}V(x) = {}_{k+1}V(x-1) + {}_{k+1}V(x+1) - {}_kV(x) + \delta V \tag{4.111a}$$

where

$$ \delta V = (\tau - 1)({}_{k+1}V(x-1) + {}_{k+1}V(x+1)) + (\rho^2 - \tau^2 + 1){}_kV(x) \tag{4.111b}$$

δV is considered as a 'perturbation' on the lossless, main part in eqn (4.111a).

The step from V^f to V^{TLM} subject to all the boundary conditions (eqn (4.104) finally consists of the adjustment of the ($k=1$)-values. In other words we must account for the difference in using eqn (4.110) for V^f and (4.104a) for V^{TLM}. This difference is given by:

$$\begin{aligned}\delta V(x,1) &= \frac{\rho+\tau}{2}(\delta(x,1)+\delta(x,-1)) - \tau(\delta(x,1)+\delta(x,-1)) \\ &= \frac{\rho-\tau}{2}(\delta(x,1)+\delta(x,-1))\end{aligned} \tag{4.112}$$

is treated as an additional source or excitation at nodes $x=-1$ and $x=+1$ at time step $k=1$ which must also be counted. As a consequence of eqn (4.110) this yields:

$$\begin{aligned}{}_kV^{\mathrm{TLM}}(x) &= {}_kV^f(x) + \frac{\rho-\tau}{2}[\,{}_{k+1}V^f(x-1) + {}_{k-1}V^f(x+1)] \\ &= \frac{\rho+\tau}{2\tau}[{}_kV^f(x) - (\rho-\tau)^2\,{}_{k-2}V^f(x)]\end{aligned} \tag{4.113}$$

This is the *exact* formula for the node voltage on any node x at any time-step k after initially exciting the node x of the network by pulses $0^iV_L(x) = 0^iV_R(x) = 1/2$.

Goldstein [4.30] has demonstrated, the manner in which this solution goes over into the fundamental solution of the telegraph equation as $k\to\infty$, $x=O(k)$. Using the asymptotic properties of the resulting modified Bessel functions Enders and de Cogan [4.10] have shown, that for large k and large $r(\equiv \tau/\rho = Z/R)$ the deviation from the fundamental solution of the corresponding diffusion equation is of the order $O(r/k)$. Since $r=\Delta t/RC$, much larger time steps are permitted, than are possible within the explicit finite difference ($R=Z$) method.

A visualisation of CRW and TLM scattering processes

There is an alternative method which can also provide the algebraic expression for the value at any node at any time. It arises out of attempts to visualise the TLM scattering process when viewed as a probabilistic process [4.32]. In theory it can be used in one, two or three-dimensions, but for anything above one-dimension it is extremely difficult to handle. The scattering parameters ρ and τ are the probabilities of reflection and transmission with respect to the current direction of motion of a pulse. We start by considering a single particle which is incident from the left on node x at time $k = 0$. Its subsequent history is shown in figure 4.8.

It can be seen that we get jumps-to-zero at every alternate time-step and that the contributions at any location constitute a binomial sequence (e.g. at $k = 2$ we have one contribution at $(x - 2)$ and one at $(x + 2)$ while we have two contributions at (x). A form of Pascal triangle is seen to build up from the point of excitation, but equally each subsequently populated point becomes an excitation point for a new Pascal triangle. Thus $(x + 1)$ at $k = 1$ will be the origin of triangle that is identical in form to the triangle excited but multiplied by τ. There will be a similar triangle originating at $(x - 1)$ at $k = 1$ but multiplied by ρ. The original triangle is in fact the superposition of these two, so that the problem could be approached recursively.

Within the scatter diagram we can also identify regions of influence and regions of dependence. Inspection of figure 4.9a will show that the shaded area falls within a triangle whose apex constitutes a limit of influence. Similarly the point identified in figure 4.9b is dependent only on the shaded region above it.

As the scattering process develops it can be seen from figure 4.8 that it fills out a binary sequence. Thus at $k = 1$ when we would expect 2^1 components we have ρ and τ. At $k = 2$, $2^k = 4$ components and we have $\rho\rho$, $\rho\tau$, $\tau\rho$ and $\tau\tau$ although their location is not obvious without inspection or without recourse to spatial shift operators [4.33]. Thus a domain of influence (figure 4.9a) can be interpreted as the origin point for a binary sequence.

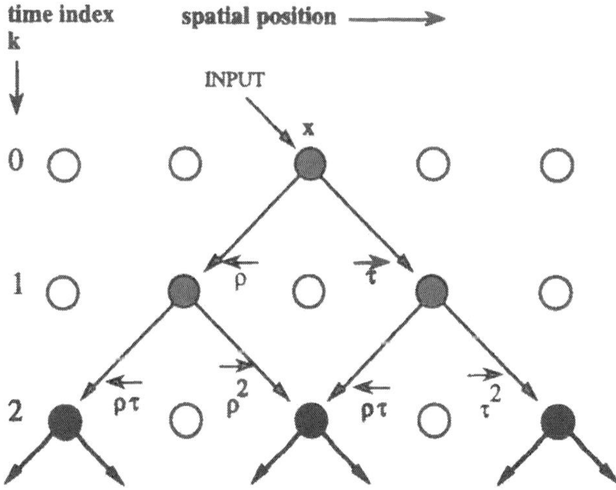

Figure 4.8 Asymmetric scatter diagram for an initial input from the left.

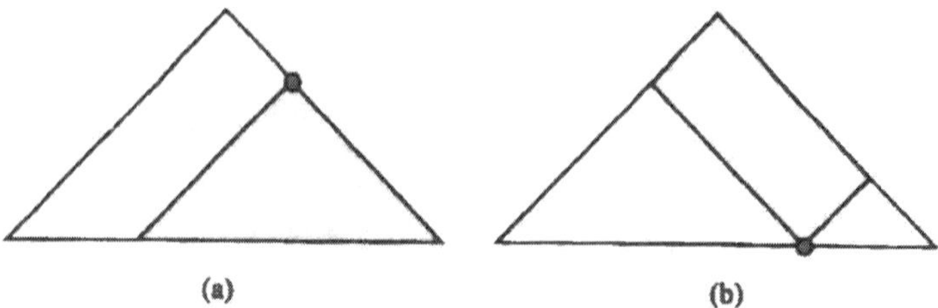

Figure 4.9 (a) One-dimensional TLM/CRW scatter diagram with domain of influence of an individual point shown shaded. (b) domain of dependence of an individual point in a scatter diagram.

Thus, if we consider the point $(x + 1)$ at $k = 1$ in figure 4.8 we can see that at $k = 2$ it will give rise to $\tau(\tau\rho)$ and $\tau(\tau)$ and at $k = 3$; $\tau(\rho\rho)$, $\tau(\rho\tau)$, $\tau(\tau\rho)$ and $\tau(\tau\tau)$. Similarly, the domain of dependence can be interpreted as all the components of the binary sequence which start at the origin and reach the destination point. This is shown in greater detail in figure 4.10 and constitutes the technique for predicting the population of any point at any time which was developed by Moravec [4.34]. In this case we are looking for the contributors at $(x + 1)$ at $k = 5$ following a single injection from the left. According to a binomial expansion there should be a total of ten. We can then plot all binary sequences of ρ,τ operations which remain within the domain of dependence (figure 4.10a) and these are shown as all possible paths in figure 4.10b. These then reduce to $5\rho^2\tau^3 + 3\rho^4\tau + \rho^3\tau^2 + \rho\tau^4$.

The Moravec approach treats this as a permutations problem based on how many ways an impulse incident from the left or from the right at the source can travel within the domain of dependence to the destination point. However, as mentioned in the previous chapter, a real input of magnitude 2 at x at $t = 0$ will scatter equally to left and right. This situation should be modelled by two super-imposed scatter diagrams starting at $k = 1$: incident from the left with origin at $(x + 1)$ and its mirror image, incident from the right with origin at $(x - 1)$. This is the derivation which will be presented below.

The domain of dependence for this derivation is shown in figure 4.11 together with the symbols which are used to delineate the frontiers of the region. It will be noted that each is defined with respect to the initial origin $(x = 0, t = 0)$. The process of computing the outcome of all combinations of scatter operators can be undertaken in order to develop an algebraic expression for the sum of all the contributors, $|J|$ at location j at iteration k. When the appropriate numerical values of ρ and τ are inserted into this expression then the results are identical with those derived from the TLM or biased random walk techniques.

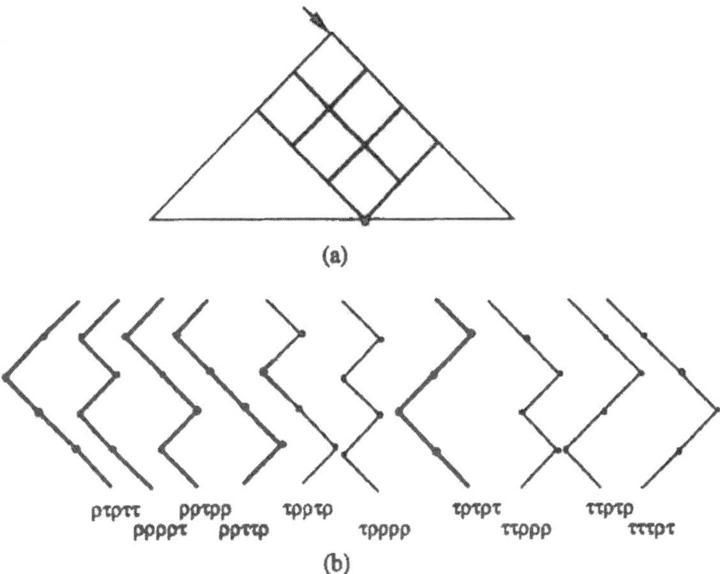

Figure 4.10 (a) The domain of dependence of the point $(x + 1)$ at time $k = 5$, (b) all possible paths which start at the origin and end at the point in question together with the reflection/transmission operations which are read from left to right.

We start by injecting a symmetrical pulse at $x = 0$ at time $t = 0$ and define the extreme points of the domain of influence in figure 4.12 as $p = k - j/2$ and $q = k + j/2$. $|J|$ then comprises the sum of all ripple paths which start at $x = 0$, $t = 0$ and terminated at $x = j$, $t = k$, while remaining within the quadrilateral domain bounded by these limits on the time diagonal and by $x = -p$ and $x = q$ on the other diagonal. Each ripple path is a series of ρ and τ operations which act on the original input. No ripple path may have a magnitude which contains an element which is greater than ρ^{2p} otherwise the path would extend beyond the domain. Therefore the solution must be of the form:

$$|J| = \sum_{n=1}^{p}[C_{n1}\rho^{2n}\tau^{k-2n} + C_{n2}\rho^{2n-1}\tau^{k-2n-1}] \tag{4.114}$$

$C_{n1} + C_{n2}$ are the iterations of all possible ripple paths for a particular value of n where C_{n1} is the number of paths which have an even number of turns and C_{n2} is the number of paths which have an odd number of turns. The proof proceeds by considering these in turn.

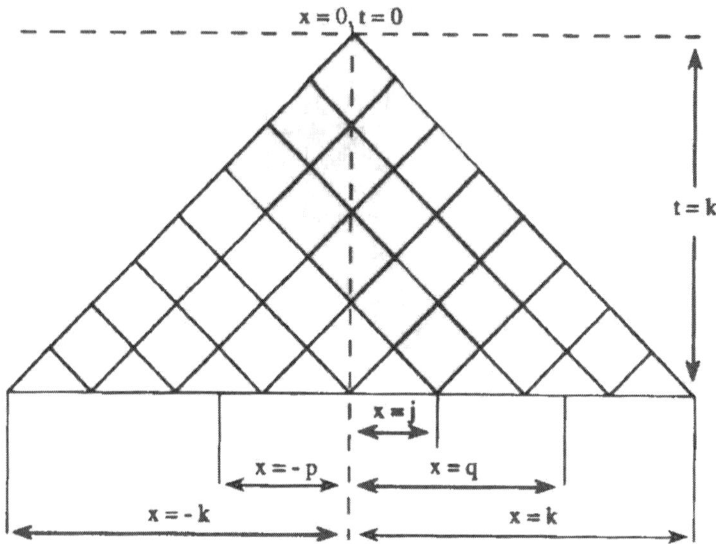

Figure 4.11 Frontiers of the domain of dependence for a symmetrical input at $x=0$, $t=0$ and a destination at $x = j$, $t = k$.

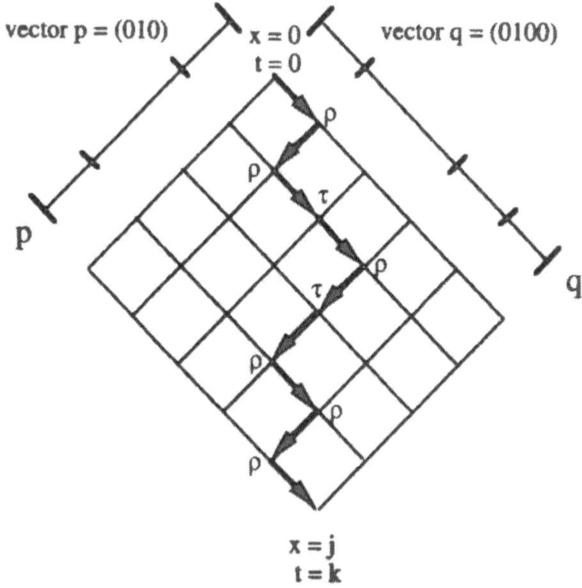

Figure 4.12 Vector mapping within the domain of dependence (even turns).

Derivation of Cn1 (even turns)

Let $C_{n1} = A_n + B_n$ where A_n are the ripple paths which begin their travel in the $+x$ direction and arrive at the terminus while moving in the $+x$ direction and B_n are the ripple paths which begin their travel in the $-x$ direction and arrive at the terminus while moving in the $-x$ direction. For a ripple path to reach position j at time k it must travel p time-steps in the $-x$ direction and q time steps in the $+x$ direction. Each transition in the direction of travel is accompanied by a ρ operation and the length of segment which is travelled in one direction is equal to the number of τ operations +1 (a length of 1 corresponds to two ρ operations in succession). The segments can be mapped onto two lines of length p and q units respectively. The turns of a particular path are represented by the number of segments allowed on each line. The length of each segment indicates the length of travel in one direction. These mappings can be represented as vectors as shown in figure 4.12.

The iterations of all ρ^{2n} ripple paths beginning in the $+x$ direction can then be counted as the number of ways in which n line segments of different non-negative integer length can be placed on a line of integer length p, times the number of ways that $n-1$ similar line segments can be placed on a line of length q. This can also be visualised as the product of the number of possible integer solutions of the equations $x_1 + x_2 + \cdots x_n = p$ and $x_1 + x_2 + \cdots + x_{n-1} = q$ for non-negative x_i.

There is a similarity with the problem of permuting n balls amongst r urns. It can be shown that there are $\binom{n-1}{n-r}$ positive integer valued solutions of the equation $x_1 + x_2 + \cdots + x_r = n$. If a vector has n values of 1 and $r-1$ values of 0 then there are $(n+r-1)!/n!(r-1)!$ permutations of this vector, where there are $(n + r - 1)$ distinct items (i.e. position of vector to be filled) and 2 distinct groups of sizes n and $r-1$.

Let x_1 be the number of 1s to the left of the first 0
Let x_2 be the number of 1s to the left of the second 0
Let x_i be the number of 1s to the left of the i-th 0

It can be seen that there is a one-to-one correspondence between each permutation of a vector and a solution of n. There are therefore $\binom{n+r-1}{n}$ non-negative valued solutions of $x_1 + x_2 + \cdots + x_r = n$. However, since each urn must already have one ball, there are only $(n-r)$ balls left for the next solution of r urns. This means that there are in fact $\binom{(n-r)+r-1}{(n-r)} = \binom{n-1}{n-r}$ solutions. If this is converted into the present problem then the number of solutions to the equation $x_1 + x_2 + \cdots + x_n = p$ is $\binom{p-1}{p-n}$ and $x_1 + x_2 + \cdots + x_{n-1} = q$ is $\binom{q-1}{q-n-1}$.

Thus $A_n = \binom{p-1}{p-n} \binom{q-1}{q-n-1}$

By means of an identical argument it can be shown that $B_n = \binom{q-1}{q-n}$ $\binom{p-1}{p-n-1}$

Derivation of C_n2 (odd turns)

To find C_{n2} we set it equal to $2C_n$, where C_n is the number of ripple paths which contain the factor ρ^{2n-1} and begin travel in the $+x$ direction and end in the $-x$ direction or vice versa. As in the previous proof we can map any ripple path on to two orthogonal lines p and q. It should be noted that in this case the segments on the left and right are the same for each direction. In other words, each ripple path can be mapped to one on the right even if p and q are not the same length (see figure 4.13). Therefore using the same form of argument as previously it can be shown that Cn is be given by $\binom{p-1}{p-n}\binom{q-1}{q-n-1}$.

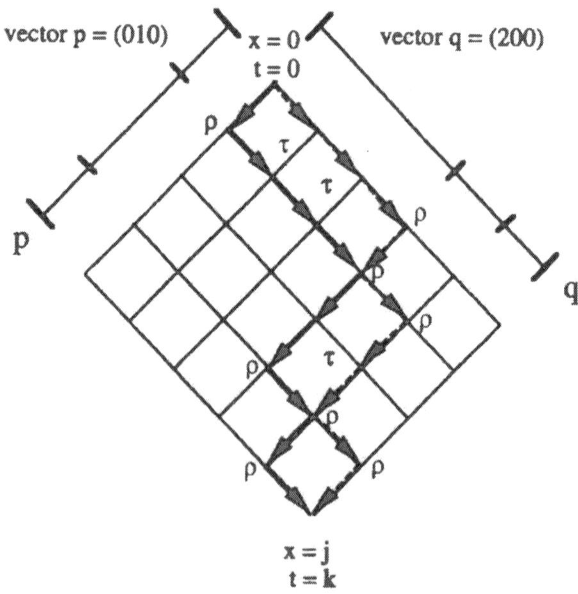

Figure 4.13 Vector mapping within the domain of dependence (odd turns).

Thus, the population $|J|$ at location j at iteration k is given by:

$$|J| = \sum_{n=1}^{p}\left[\left[\binom{p-1}{p-n}\binom{q-1}{q-n-1} + \binom{p-1}{p-n-1}\binom{q-1}{q-n}\right]\rho^{2n}\tau^{k-2n-2} + 2\left[\binom{q-1}{q-n}\binom{p-1}{p-n}\right]\rho^{2n-1}\tau^{k-2n-1}\right] \tag{4.115}$$

The above arguments can also be applied to unidirectional injection of carriers. For a pulse which is incident from the right at $x = 0$, $t = 0$ it can be shown that

$$|J'| = \sum_{n=1}^{p}[(\tau A_n + \rho B_n)\rho^{2n}\tau^{k'-2n-2} + (\rho + \tau)C_{n2}\rho^{2n-1}\tau^{k'-2n-1}] \tag{4.116}$$

where $k' = k + 1$ and $j' = j + 1$.

The magnitude at position j at time k due to a single impulse from the left at $x = 0$, $t = 0$ is $|J''|$ and the expression is identical in form to eqn. (4.116) with $k'' = k-1$ and $j'' = j-1$. It can be easily shown that $|J'| + |J''| = |J|$.

The practical value of Eqn (4.115) can be demonstrated by calculating the population at $j=3$ following a symmetrical input at $x=0$, $t=0$ which has been allowed to progress for 7 iterations. In this case $p=2$ and $q=5$.

Accordingly $|J| = [A_1+B_1]\rho^2\tau^4 + 2C_1\rho\tau^5 + [A_2+B_2]\rho^4\tau^2 + 2C_2\rho^3\tau^3$

$$= \left[\binom{4}{3}\binom{1}{0} + \binom{1}{0}\binom{4}{4}\right]\rho^2\tau^4 + 2\binom{4}{4}\binom{1}{1}\rho\tau^5 + \left[\binom{4}{0}\binom{1}{2} + \binom{1}{-1}\binom{4}{3}\rho^4\tau^2 + 2\binom{4}{3}\binom{1}{0}\right]\rho^3\tau^3$$

$= 5\rho^2\tau^4 + 2\rho\tau^5 + 6\rho^4\tau^2 + 8\rho^3\tau^3$ which is what would be obtained if the TLM process were undertaken symbolically using Mathematica or Maple. When suitable values of ρ and τ are substituted then the value is identical to the analytical results for a Gaussian diffusion.

A statistical view of lossy TLM

The scatter diagrams of the previous section were first developed in an effort to analyse the effects of mesh truncation. It was believed that significant savings in computational time could be achieved by arranging for the mesh in a lossy TLM problem to be only as big as the problem being modelled [4.35]. So, for instance, in the modelling of heat-flow in a metal rod following a single-shot injection we could have the TLM network expand in advance of the diffusing perturbation. The effects of mesh truncation were investigated by comparing the difference in the terms resulting from the scatter diagrams for an open and a truncated mesh with a boundary reflection, $\rho_{\mathbf{boundary}} = 0$.

The alternative approach is to view the matter as a statistical problem. A single-shot injection will lead to a Gaussian diffusion which will have a standard deviation, σ, which depends on ρ, τ and k. The error due to the truncation of a Gaussian distribution at a given number of standard deviations is well known and it was felt that the relationship between σ, and the TLM parameters ρ,τ and k would constitute a better basis for mesh truncation.

This was approached using statistical analysis by Cox [4.36]. The first step involved an investigation of the conditions when the TLM simulation of single shot injection could be considered as a Gaussian distribution. Analysis used a one-dimensional bar of material where the space and time discretisations were arranged so that the TLM model depended only on the reflection coefficient, ρ. A single pulse of magnitude equivalent to a temperature rise of 200°C was injected into the centre so that heat diffused symmetrically. Temperature profiles were collected after every iteration and were tested for normality using the Chi-square test. Figure 4.14 shows the Chi-square test probability plotted against iteration number, k for different values of ρ. If we accept a 95% confidence limit for the test, we may say that a profile is a normal distribution when the ordinate value in the figure is above 0.95. Hence, when $\rho = 0.5$, the distribution is normal after only 3 iterations but for $\rho = 0.1$, approximately 27 iterations are required before a normal distribution is reached. These tests showed no difference for profiles calculated using either link-resistor or link-line nodal configuration.

The profile data was used to calculate variance (σ^2) which was then plotted against k (after convergence to normality) for different values of ρ. The dependence of the variance on k was found to be linear in all cases and straight lines were fitted to the points using a standard linear regression routine. The regression lines have the form $ak + b$ and 95% confidence intervals for the values of a were calculated. The largest confidence intervals was 1.7×10^{-5} for $\rho = 0.1$. For $\rho = 0.5$, the confidence interval was zero i.e. the linear fit was perfect to within the numerical accuracy of the computer. Analysis of the dependence of σ^2 on ρ for the link-line node showed that:

$$_{LL}\sigma^2(k) = \frac{\tau}{\rho}k + \frac{1}{2} - \frac{\tau}{\rho}\left[1 - \frac{\tau}{2\rho}\right] \tag{4.117}$$

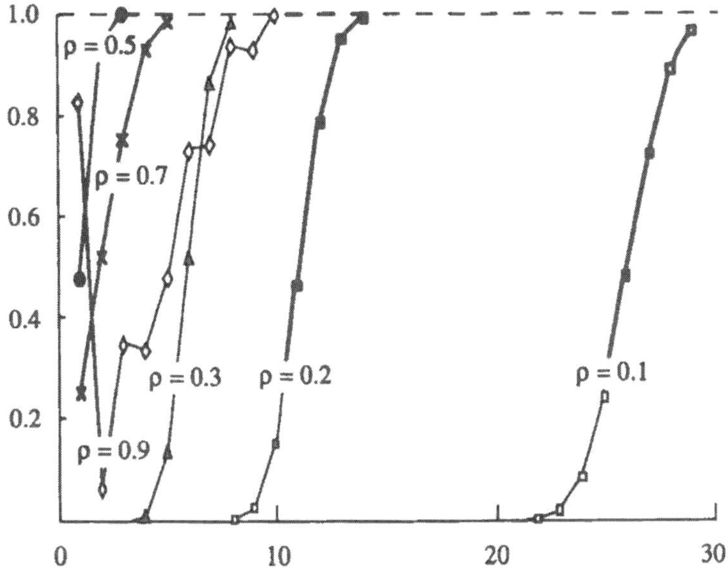

Figure 4.14 Signicance of fit to a Gaussian distribution (vertical axis) versus number of iterations (horizontal axis) as a function of ρ for a single injection into an infinite one-dimensional sample.

A similar analysis for the link resistor node showed that it is smaller by 0.5

$$_{LR}\,\sigma^2(k) = \frac{\tau}{\rho}k - \frac{\tau}{\rho}\left[1 - \frac{\tau}{2\rho}\right] \tag{4.118}$$

These results provide a useful metric in dynamically expanding diffusion models.

Conclusions

This chapter has attempted to cover the current state of the theory of lossy TLM. It should now be quite clear that the processes can be treated in a probabilistic way and are amenable to statistical analysis and algebraic manipulation. These perspectives allow us to use a wide range of techniques which would not be obvious if lossy TLM were interpreted strictly in electromagnetic terms. While the treatment has attempted to be wide-ranging there is no claim that it has covered all aspects. Indeed it was felt better to defer discussion on several topics to the relevant applications chapters which follow.

References

[4.1] P.B. Johns, A simple, explicit and unconditionally stable routine for the solution of the diffusion equation, *Int. Jnl Numerical Methods in Engineering*, 11 (1977) 1307-328.

[4.2] C.P. Kenny, TLM and finite difference; an algebraic comparison, *PhD thesis*, University of East Anglia, Norwich 1997.

[4.3] G. Lancaster, Introduction to Fields and Circuits OUP 1992, pp. 268-273.

[4.4] G. Lancaster Introduction to Fields and Circuits OUP 1992, pp. 348-349.

[4.5] A.E. Kennelly, Tables of complex hyperbolic and circular functions Harvard University Press 1921.

[4.6] K.L. Moravec, Transmission Line Matrix modelling: using the propagation constant as a measure of error *Part I of BSc thesis*, UEA Norwich 1994.

[4.7] E.C. du-Fort and S.P. Frankel, Stability conditions in the numerical treatment of parabolic differential equations. *Math. Tables and Other Aids to Comp.*, 47 (1953) 1135.

[4.8] S.H. Pulko, A. Mallik, R. Allen and P.B. Johns, Automatic timestepping in TLM routines for the modelling of thermal diffusion processes *Int. Jnl. of Numerical Modelling* 3 (1990) 127-136.

[4.9] X. Gui, P.W. Webb and D. de Cogan, An Error Parameter in TLM Diffusion Modelling, *Int. Jnl. of Numerical Modelling*, 5, 129-137, (1992).

[4.10] P. Enders and D. de Cogan, The efficiency of transmission line matrix modelling-a rigorous viewpoint, *Int. Jnl. of Numerical Modelling*, 6 (1993) 109-126.

[4.11] A.J. Lowery, A study of the static and multi-gigabit dynamic effects of gain spectra carrier dependence in semiconductor lasers using a transmission line laser model, *IEEE Jnl. of Quantum Electronics*, 24 (1989) 2376-2385.

[4.12] S.H. Pulko and C.P. Phizacklea, Modelling of mass transfer and related phenomena by the TLM technique, *Proceedings of MIC'89*, Grindlewald (1989).

[4.15] I.A.D. Russell and P.W. Webb, Simulation of electro-thermal instability in microwave devices using the TLM method in "TLM: The wider applications" (Ed. D. de Cogan), School of Information Systems (UEA) 1996 pp 4.1-4.10.

[4.16] A. Soulos, D. de Cogan and P. Enders, TLM: order of accuracy enhancement, *Proceedings of the Second International Workshop on Discrete Time Domain Modelling of Electromagnetic Fields and Networks*, Hotel Ambassador. Berlin, 28-29 October 1993.

[4.17] J.S. Nielsen and W.J.R. Hoefer, Generalised dispersion analysis and spurious modes of 2-d and 3-d TLM formulations, *IEEE Trans. on Microwave Theory and Techniques* MTT-41 (1993) 1375-1384 (note: a typographical error on page 1375 of this reference gives the volume number as 11).

[4.18] M. Krumpholz, Dispersion characteristics of two and three dimensional TLM schemes *Proc. First International Workshop on Transmission Line Matrix (TLM) Modelling - Theory and Applications*, University of Victoria (BC), August 1995 pp 27-36.

[4.19] P. Russer, On the field theoretical foundations of the transmission line matrix method, *Proc. First International Workshop on Transmission Line Matrix (TLM) Modelling - Theory and Applications*, University of Victoria (BC), August 1995 pp 3-12.

[4.20] M.I. Sobhy, The order of complexity and degree of freedom of lumped commensurate and non-commensurate networks, *Int. Jnl. of Electronics*, 50 (1981) 47-54.

[4.21] S.H. Pulko, A.J. Wilkinson and M. Gallagher, Redundancy and its implications in TLM diffusion models, *Int. Jnl Numerical Modelling*, 6 (1993) 135-144.

[4.22] X. Gui and P.W. Webb A comparative study of two TLM networks for the modelling of diffusion processes *Int. Jnl Numerical Modelling*, 6 (1993) 161–164.

[4.23] R. Ait-Sadi and P. Naylor, An investigation of the different TLM configurations used in the modelling of diffusion problems, *Int. Jnl Numerical Modelling*, 6 (1993) 253–268.

[4.24] P. Enders and D. de Cogan, Discrete Modelling of Transport Processes in Two Spatial Dimensions, *Int. Jnl. of Numerical Modelling*, 5 (1992) 121–129.

[4.25] C.L. Phillips and H.T. Nagle, "Digital control system analysis and design" Prentice-Hall, 1995, pp. 53–77.

[4.26] J.E. Marshall, "Control of time delay systems" Peter Peregrinus, 1979, pp 131–146.

[4.27] A.R.M. Witwit, A.J. Wilkinson and S.H. Pulko, A method for algebraic analysis of the TLM algorithm, *Int. Jnl. of Numerical Modelling*, 8 (1995) 61–71.

[4.28] D. de Cogan and P. Enders, Discrete Green's functions and hybrid modelling of thermal and particle diffusion *Int. Jnl. of Numerical Modelling*, 7 (1994) 407–418.

[4.29] Y. Taitel, On the parabolic, hyperbolic and discrete formulation of the heat conduction equation *Int. J. Heat and Mass Transfer* 15 (1972) 1369–371.

[4.30] S. Goldstein, On diffusion by discontinuous movements, and on the telegraph equation, *Quart. Jnl. Mech. and Appl. Math.*, 1. IV, (1951) 129–156.

[4.31] P. Enders and D. de Cogan, Discrete models of heat-flow in layered materials using simple and correlated random walks, *Int. Jnl. of Numerical Modelling* 9 (1996) 445–457.

[4.32] D. de Cogan and P. Enders, Microscopic effects in TLM heat-flow modelling *IEE Colloquium Digest* 1991/192, 8.1–8.11.

[4.33] D. de Cogan and A. de Cogan, Numerical Modelling for Engineers Oxford University Press 1997.

[4.34] K.L. Moravec, Transmission Line Matrix modelling: an exact solution for a single pulse input, *BSc thesis*, (Part II), UEA Norwich, 1994.

[4.35] A. Smith, Transmission Line Matrix optimisation and application to adsorption phenomena, *BSc thesis*, Nottingham University 1988.

[4.36] S.J. Cox (unpublished work).

Chapter 5

Heat transfer applications of lossy TLM algorithms

Introduction

Johns initiated the application of TLM to thermal diffusion in 1977 [5.1]. He revisited the subject in 1979 [5.2], but thereafter there was little work done until four years later. Since 1982 there have been many papers which address the subject in one form or another. Most of these have been concerned with conductive heat transfer. Convective and radiative processes have really only been considered as lossy boundaries or as heat sources in conduction problems.

The first TLM thermal models were simple affairs, but they very quickly developed to include non-linear effects. The extension to larger problems, numerically stiff problems and problems with complex geometries required variable meshes and variable time-steps. The development of these models has led to an increased understanding of the power of TLM and it has often provided fresh insights into the nature of the problem being modelled. Coupled TLM models have followed and still present much scope for innovation.

This chapter starts by presenting some tutorial details on the transition between material parameters and a TLM algorithm. In so far as is possible, it then goes on to treat various topics in the order of their evolution and considers practical applications and outcomes on the way.

General theory

The thermal capacitance a material is given by

$$\begin{aligned} C &= \text{(density) (specific heat) (volume)} \\ &= \rho C_p A_{yz} \Delta x \end{aligned} \tag{5.1}$$

where A_{yz} = area and Δx is the discretised length in the x direction. In this case ρ is the conventional physics symbol for density and C_p is the specific heat at constant pressure. It is usual to assume that $A=1$ for one-dimensional problems so that the distributed capacitance, $C_d = \rho C_p$.

In two- and three-dimensional formulations it is normal but not essential to have identical spatial discretisations in all coordinate directions. The main question is how the capacitance will be shared amongst the different transmission lines. Capacitance is related to line impedance through the time discretisation:

$$Z = \frac{C}{\Delta t} \tag{5.2}$$

We can see in figure 5.1 that the capacitance can be split into two parallel parts and therefore into two parallel impedances. This means that:

$$Z_x = Z_y = \frac{C}{2\Delta t} \tag{5.3}$$

It is for this reason that some authors prefer to start by defining eqn (5.1) as: aC = (density) (specific heat) (volume), where $a = 1$, 2, 3 depending on whether we are working in one-two or three dimensions. If this is used then the individual line impedances are defined as in eqn (5.2).

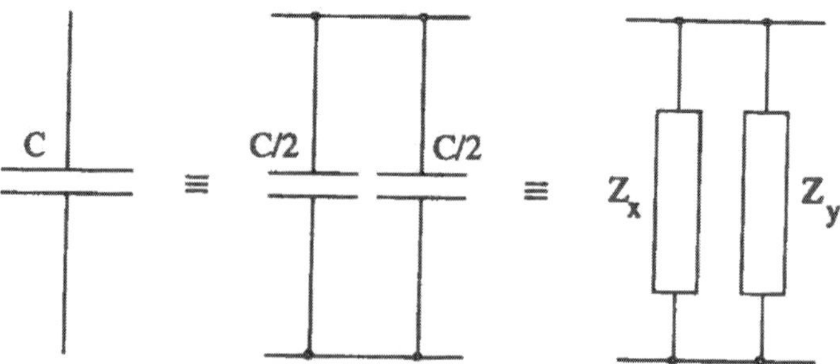

Figure 5.1 The division of a capacitance between coordinate directions in two-dimensions.

It was mentioned above that there is no absolute requirement to have $\Delta x = \Delta y = \Delta z$. However within TLM there is a requirement to have pulse synchronism. Thus, we can, if essential, subdivide the nodal space so that the capacitances and thereby the impedances are different, but sharing a common Δt.

There are similar considerations for nodal resistance, but there is no influence by Δt. Instead, the resistance is affected by the material thermal conductivity and the local geometry. Using the electrical analogue, we have:

$$R = \frac{\text{(contact separation)}}{\text{(contact area) (thermal conductivity)}} \tag{5.4}$$

We define the 'contact separation' as the discretisations Δx, Δy and Δz shown in figure 5.2 and 'contact area' as A_{xy}, A_{yz}, A_{zx} depending on the direction of heat-flow.

In a three-dimensional TLM node there are effectively six resistors, so that the total resistance in any coordinate direction must be split into two parts and placed one on either side of the node centre. So long as the node is symmetrical then this does not present a problem. The approach to non-uniform nodes was outlined in chapter 3. We will now see how a polar node like that shown in figure 5.3 is converted into an equivalent network. If we assume a unit height then the area is dependent only on the arc-length. The arc denoted by the numbers (1,2) which runs through the node centre might

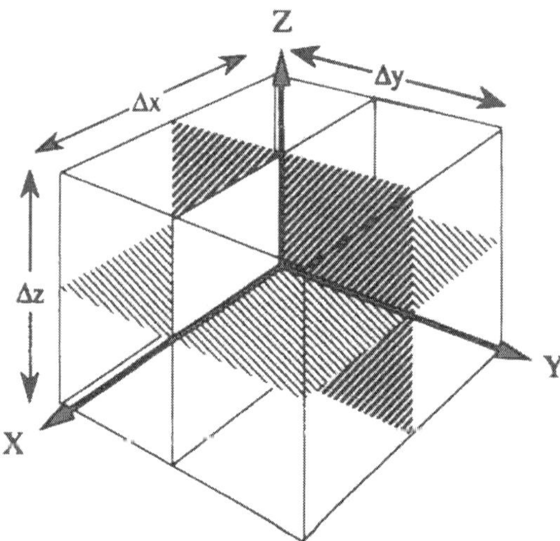

Figure 5.2 A cubic node showing the discretisations and the planes representing areas A_{xy}, A_{yz}, A_{zx}.

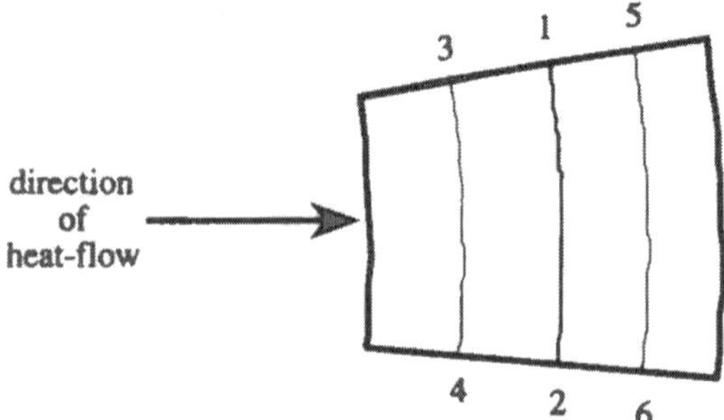

Figure 5.3 'Area' descriptions for thermal resistance in a polar node.

seem to be a good mean value for a link-line network. Others might feel that a more accurate description might involve two different resistors, one to the left and one to right of the centre, having areas described by the arc-lengths (3, 4) and (5, 6). Of course in a link-resistor TLM network the inner and outer faces of the node might be the more appropriate areas.

Simple TLM heat-flow algorithms

In 1982 de Cogan and John [5.3] published their initial results on using TLM to study the thermal behaviour of a form of electrical transient overload suppresser (the punch-through diode). There was no clear understanding of the difference between different nodal descriptions at that time so that both seem to have been used interchangeably in the further work in this area which appeared in 1985. These included two-dimensional treatments [5.4], a fairly elementary non-linear model for the punch-through diode with position dependent power dissipation [5.5] and a model for a microwave diode soldered into a package [5.6].

Models with complex geometries

When de Cogan and John [5.4] developed 2D TLM models to treat cylindrical geometries of the form shown in figure 5.4 they observed anomalous cooling at the centre of the device. That this was an artefact was easily confirmed by using sets of progressively finer meshes. The source of this problem was found to be an early misunderstanding as to what was the conserved entity in such models [5.7]. If we use the analogues $T \equiv V$ and Heat $\equiv$ I then it is current that is conserved during the reflection shown in figure 5.5.

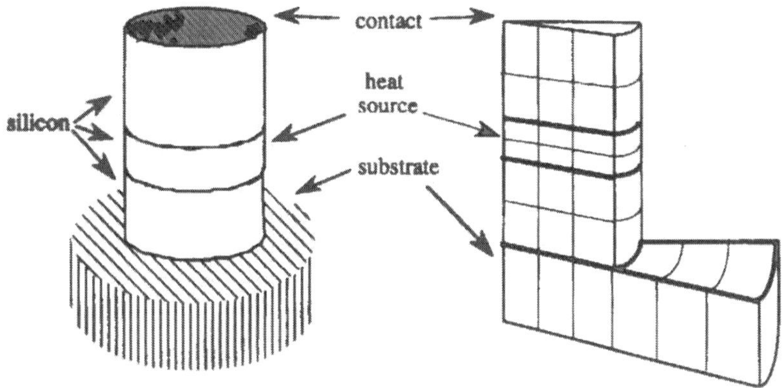

Figure 5.4 A cylindrical diode with heat source and its two-dimensional mesh implementation.

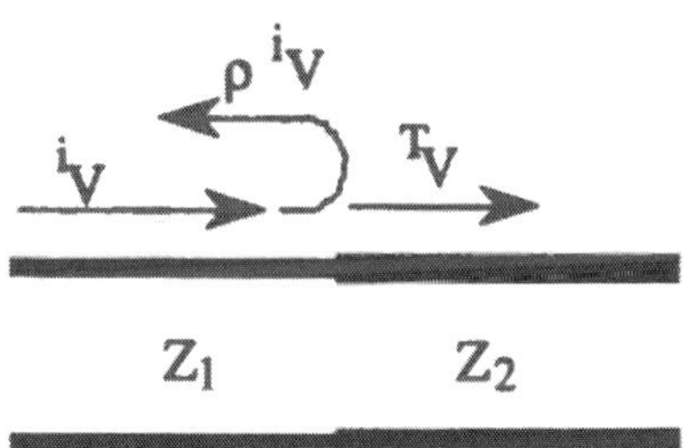

Figure 5.5 Reflection/transmission of a pulse which is incident at an impedance mismatch.

Since incident current = reflected current + transmitted current we have:

$$\frac{{}^{i}V}{Z_1} = \rho\frac{{}^{i}V}{Z_1} + \frac{{}^{T}V}{Z_2} \tag{5.5}$$

where $\rho = (Z_2 - Z_1)/(Z_2 + Z_1)$

So, while the transmitted current ${}^{T}V/Z_2 = (1-\rho)({}^{i}V/Z_1)$

The transmitted voltage is given by:

$${}^{T}V = (1-\rho)\,\frac{Z_2}{Z_1}\,{}^{i}V = (1+\rho){}^{i}V \tag{5.6}$$

In 1985, while working with Peter Johns, Pulko developed the first techniques to treat a truly complex geometry. Their subject was a two-dimensional model of a jet engine turbine blade. Heat is absorbed by the blade as the combustion gases pass. This must be transferred down the blade (see figure 5.6) to the shaft, which acts as a heat sink. It was reported that temperature dependent thermal parameters were used in the model [5.8]. The details of achieving a variable meshing were not included but the influence of the inductance term ($L = \Delta t^2/C$) was discussed.

Reference [5.6] gives more details of the process of achieving synchronism in a variable mesh. Figure 5.7 shows two adjacent nodes of the same material with different dimensions. Pulses traversing nodes A and B must do so in exactly the same unit of time, Δt.

If the impedance of A is given by:

$$Z_A = \frac{\Delta t}{C_d \Delta x} \tag{5.7}$$

then the impedance of B is given by:

$$Z_B = \frac{\Delta t}{C_d b \Delta x} = \frac{1}{b} Z_A \tag{5.8}$$

It is interesting to note that some time later this process was used in reverse to obtain matching between adjacent materials of different thermal properties [5.9].

There is some uncertainty as to the priority on the use of stubs in thermal conduction modelling as it is mentioned in references [5.6–5.8].

Details of the first three-dimensional models appeared in print in 1987. All included variable meshing and temperature dependence of thermal

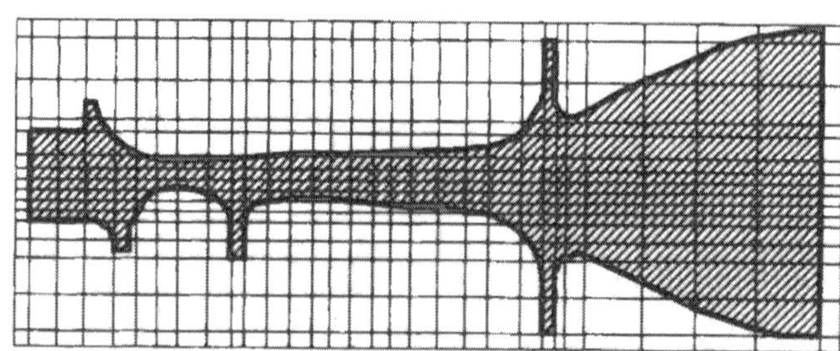

Figure 5.6 The variable meshing along the length of a jet engine turbine blade.

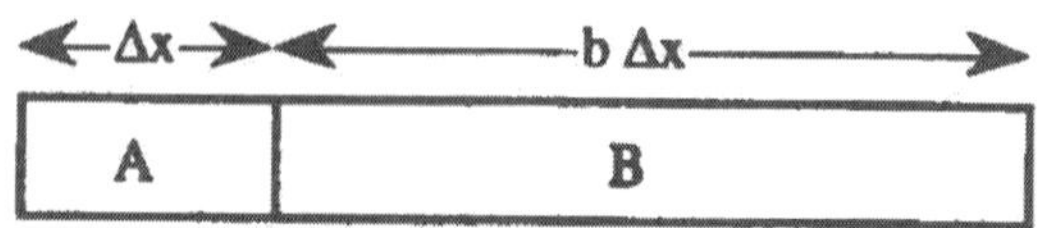

Figure 5.7 Adjacent nodes in a variable mesh configuration.

parameters [5.10, 5.11]. The latter reference was probably the first case of replacing a node by a lumped circuit element. The problem (a substrate electric fuse) comprised a thin silver layer deposited on a ceramic substrate. Current passing through the silver generated heat at a geometrical constriction in the electrical conduction path. At first the silver was assumed to be so thin that it acted only as a heat source and did not provide a thermal conduction path. In this realisation it was not possible to obtain agreement between the simulated results and experimental data obtained using infra-red thermal imaging. Once it was realised that the silver layer did contribute significantly to the thermal behaviour the next problem was how to include it in the model. A mesh dimension equivalent to the silver ($\sim 1\mu$m) would have increased the computational load unacceptably. Instead it was decided to increase heat conduction on the surface by including silver resistances: $R_{Ag} = 0.5\Delta x/\{(k_T)(\text{thickness})(\text{width})\}$, where k_T, thickness and width refer to the silver film. These were placed in parallel with the lateral resistances of the substrate nodes next to the surface immediately below the silver. The additional thermal mass due to the silver was negligible. The outcome of this alteration was a significant improvement in the level of agreement between theory and experiment.

In 1993 there were two papers which attempted to resolve the question of whether a link-line or a link resistor formulation was the more appropriate. For several reasons they appear to have clouded the issue. The technique of separating two nodes by a section of transmission line has a long tradition of being referred to as a 'link-line' within the TLM community. It is for that reason that the expressions 'link-line' and 'link-resistor' have been used throughout this book. Both of these papers used alternative, non-conventional nomenclatures. Secondly, it is felt that the order of publication in the journal was unfortunate. The 'RB' node of Ait-Sadi and Naylor [5.12] and the 'modified' node of Webb and Gui [5.13] are both link-resistor nodes. The latter authors who were published first push the link-resistor node to the extreme of $Z/R = 10$ (equivalent to $\rho = 0.099$) and note that it then displays oscillations following impulsive excitation. In the same experiment on a link-line node their results show a step-wise decrease in temperature. They do not say whether they take account of the fact that the link-resistor node is displaced in space by $\Delta x/2$ and in time by $\Delta t/2$. There is a suspicion that the steps in their link-line results are simply the mid-points between the extrema in a damped wave. However, they do highlight problems associated with time-step changes during the modelling of a transient. Ait-Sadi and Naylor [5.12] provide a comprehensive 'how-to' comparison of the two nodal descriptions and clearly recognise that effective comparisons need to be between the same spatial location at the same instant in time. They comment on problems associated with boundary treatments in link-resistor models.

Complex geometries may also require spatial sub-structuring: finer meshes in some regions. Although this was described in detail in chapter 3 it is worth noting that the contribution of Pulko *et al* in this area was for thermal modelling of the process of glassware manufacture [5.14]. Further details on the application of TLM in glass pressing are presented by Pulko and Phizacklea [5.15].

Witwit [5.16] has contributed to the subject of modelling heat-flow in complex geometries using non-conventional spatial discretisations. These include triangular and hexagonal meshes. His approach has already found a use in anisotropic heat-flow in mono-directional carbon fibre composite material [5.17]. This is believed to be the first case where TLM has been used to model a heat-flow equation which contains cross-derivatives:

$$\rho C_p \frac{\partial T}{\partial t} = k_{\parallel} \frac{\partial^2 T}{\partial x^2} + k_{\perp} \frac{\partial^2 T}{\partial y^2} + (k_{\parallel} + k_{\perp}) \frac{\partial^2 T}{\partial x \partial y} \qquad (5.9)$$

(where $k_{\parallel}$ and $k_{\perp}$ are the parallel and perpendicular thermal conductivities).

Non-linear models

The specific heat and thermal conductivity of most materials are both functions of temperature. For many materials near and above room temperature the variation of specific heat is very slight and can often be ignored. This is not the case with thermal conductivity which can vary significantly. Similarly if a crystalline material is being modelled there may also be an anisotropy.

The effect of solder thickness and solder voids on the thermal performance of power semiconductors was studied [5.18] using a two-dimensional model similar to that in figure 5.4. Fixed-point thermal data were stored in the programme. At the end of each iteration $k_T({}_kT(x,r))$ and $C_p({}_kT(x,r))$ for the next time-step were obtained by interpolation using these values. Only in the case of very rapid transients in situations where there is a significant variation in thermal parameters was it found necessary to estimate these using ${}_{k+1}T(x,r)$ rather than ${}_kT(x,r)$ [5.19]. In this situation an iteration within iteration was used: a first estimate of ${}_{k+1}T(x,r)$ was made using $k_T({}_kT(x,r))$ and $C_p({}_kT(x,r))$. This was then fed back to obtain revised values of k_T and C_p so that a new estimate of ${}_{k+1}T(x,r)$ could be made. The process was repeated until convergence was obtained (usually about 3 cycles).

The non-linear behaviour of thin-film fuses on substrates helped to identify links to some interesting physical phenomena. Fused silica (SiO_2)

becomes a better thermal conductor as it gets hotter. Thus a negative thermal feedback mechanism can operate. The variation of maximum temperature as a function of time is linear and is consistent with I^2t behaviour which is observed in real electric fuses which are embedded in sand. This is in contrast to the $I^2\sqrt{t}$ behaviour which is observed in silicon devices under electrical overload conditions. The thermal conductivity of silicon decreases with increasing temperature. The effects of positive thermal feedback can be observed in the transient thermal failure of metal film resistors [5.20]. It is also predicted that thin-film fuses on beryllia substrates should display a very sharp transition between thermal stability and failure [5.21], although this has yet to be experimentally confirmed.

Not all non-linear models are so definite in their outcomes. A recent study of the WRITE process in magneto-optic (MO) disks [5.22] was unable to obtain agreement with experimental data using published results. In the first instance there is no currently available data on the temperature dependence of the thermal conductivity of gadolinium-terbium-iron films. The authors found that in order to approach the experimental data they had to adopt a thermal conductivity which was ten times smaller than the published bulk value. Increased agreement was obtained when a temperature dependence of this reduced thermal conductivity was included in the model. This apparent discrepancy is now attributed to the non-ideality of the interface between the film and the substrate [5.23].

Automatic time-step changes

Johns was the originator of the concept of time-step changes as a means of increasing computational efficiency. He identified the need for an error parameter which could be used as a means of triggering a time-step change automatically within a simulation. The ideas which are outlined in chapter 3 were brought together in the paper of Pulko *et al* [5.24].

Webb and Gui [5.25] have extended the treatment to three dimensional problems. They present a comprehensive analysis and apply their adaptation to model the heat-flow in a gallium arsenide microwave power transistor.

Webb and Russell [5.26] have recently developed a novel measure of the inductance error which can be used with automatic time-stepping and was discussed in detail in the previous chapter. This operates on the principle that the rate of change of current through the transmission line gives rise to a voltage drop across the line inductance. This can either be included within the estimate or the time-step can be altered so as to minimise this error at all time.

Boundaries and interfaces

Almost all of the early TLM models assumed perfect heat transfer between materials and well defined boundaries. However, the turbine blade which was modelled by Pulko *et al* [5.8] had to include heat transfer between a hot moving gas and a conductive medium. The goal of a TLM algorithm for heat transfer by forced convection was achieved by Pulko *et al* [5.27].

(i) Convective boundaries

The mean temperature for a fluid moving over a heat source is

$$\frac{\rho_g C_v A_c}{\text{Per}} \frac{\partial T_g}{\partial t} + \frac{A_F C_p}{\text{Per}} \frac{\partial T_g}{\partial x} + h(T_{adw} - T_s) = 0 \tag{5.10}$$

'Per' is a periphery, h is a heat transfer coefficient, T_s = temperature of solid and T_{adw} is the adiabatic wall temperature which as an approximation is taken equal to the mean fluid temperature.

For the case of still fluid the second term in this equation is zero so that we have:

$$\frac{\rho_g C_v A_c}{\text{Per}} \frac{\partial T_g}{\partial t} = -h(T_f - T_s) \tag{5.11}$$

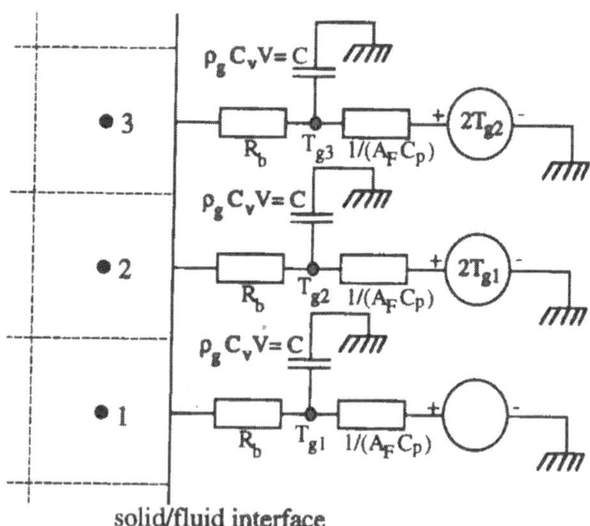

Figure 5.8 An electrical network representation of convective heat transfer.

If on the other hand the fluid velocity is high then the first term is negligible and we then have:

$$\frac{A_F C_p}{\text{Per}} \frac{\partial T_g}{\partial x} = h(T_f - T_s) \tag{5.12}$$

or

$$A_F C_p \Delta T_g = h(T_g - T_s) \text{ Per } \Delta x \tag{5.13}$$

($T_g = T_f$ if the fluid is a gas)

In TLM terms this can be modelled by having resistors of value $(A_F C_p)^{-1}$ as shown in figure 5.8. We then have a current generator/resistor combination which can be modelled by means of a suitably charged capacitor which is connected sequentially to each of the boundary nodes (i.e. it is connected to boundary node 1 for a time and then removed to boundary node 2 and so on).

Thus, we have charge passed through resistor = charge lost by capacitor

$$h(T_g - T_s) \text{ Per } \Delta x = C_p M(T_{g1} - T_{g2}) \tag{5.14}$$

The current passing through the boundary resistor, R_b is $C_p M(T_{g1} - T_{g2})/\Delta x$.

The heat reservoir may be represented by a capacitor of value $C_p\ A_g\ \Delta t$.

The other components in the electrical network of figure 5.8 are $R_b = A/h$ and a stationary capacitor ($\rho_p\ C_v\ V$) which represents the heat storage in the gas while it is resident in the vicinity of a surface node interface.

These electrical elements can now be incorporated into a TLM representation for forced convective heat transfer (see figure 5.9). The capacitors are replaced by open-circuit half length stubs.

The impedance of the stationary stub is given by:

$$Z_{\text{stationary}} = \frac{\Delta t}{2\rho_g C_v V} \tag{5.15}$$

The mobile stubs model changes in gas temperature as it moves over the surface of the solid. Each mobile stub is connected to each boundary node in turn and its impedance is:

$$Z_{\text{mobile}} = \frac{\Delta t}{2C_p A_g \Delta t} = \frac{1}{2C_p A_g} \tag{5.16}$$

In their paper [5.27] Pulko *et al* apply this formulation to a hot gas moving over the surface of a metal disk and provide comparisons with a finite element simulation.

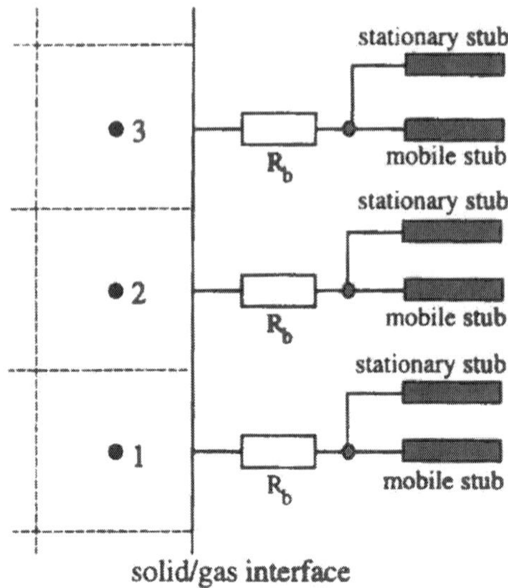

Figure 5.9 The TLM implementation of a forced convection model.

Pulko in her work has tended to include non-ideal heat transfer as a matter of course. The model for cyclic glass pressing [5.15] includes a contribution for the transfer coefficient at the glass-mould interface.

(ii) Radiative boundaries

Radiative heat transfer is another possible boundary/interface in thermal conduction models. Early (unpublished) attempts by de Cogan to treat heat-loss or uptake by radiation suggested that there were considerable problems if a conventional link-line node were used. The approach which is presented here is for a link-resistor node. If T is the surface temperature in Kelvin then the heat loss from a black-body radiating surface is given by the Stefan-Boltzman law:

$$\text{Heat loss per second per unit area} = \sigma T^4 \tag{5.17}$$

In order to estimate the heat-loss, we must have a precise measure of the surface temperature. The temperature of the nearest node will not do as it is displaced from the surface both in space and time. de Cogan [5.28] analysed the system as a TLM network by treating the heat loss as a current (designated I_{RAD}) flowing through a non-linear resistor (designated R_{RAD}) as shown in figure 5.10.

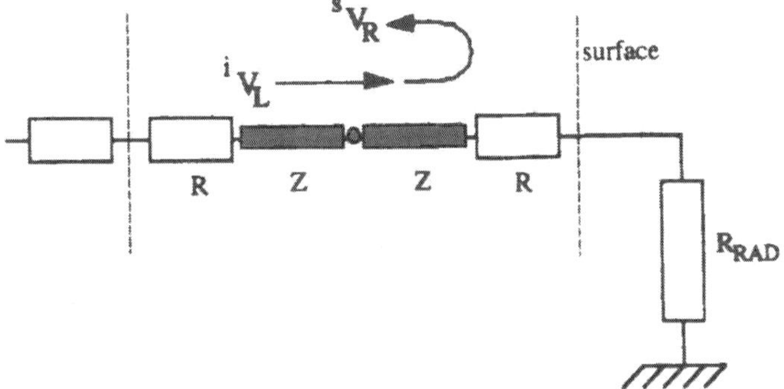

Figure 5.10 A link-resistor node with a non-linear (radiating) resistor.

The pulse approaching the nodal resistor at the surface is given by:

$$ {}_k^iI = \frac{{}_k^iV_L}{Z} \tag{5.19} $$

The current pulse reflected from the surface is:

$$ {}_k^rI = \frac{{}_k^iV_L}{Z} - I_{\mathrm{RAD}} \tag{5.20} $$

Thus at the instant that the pulse arrives at the surface we have the superposition of the approaching and reflected. The surface temperature is then given by:

$$ \begin{aligned} V_{\mathrm{surface}} &= 2\,{}_k^iV_L - ZI_{\mathrm{RAD}} \\ &= 2\,{}_k^iV_L - Z_\sigma(V^4_{\mathrm{surface}} - V^4_{\mathrm{ambient}}) \end{aligned} \tag{5.21} $$

This is a quartic equation which will have to be solved iteratively at every time-step. Once we have I_{RAD} we then have the pulse scattered at the surface resistor which becomes incident on the node centre (from the right) at the next iteration.

In drawing this section to a close we note that there has not yet been an attempt to devise a TLM model of free convective heat transfer, although the developments by Witwit *et al.* [5.17] and by Langley *et al* [5.29] may soon change that situation.

Coupled models

The foregoing sections have outlined the development of TLM models for heat-flow applications. We have witnessed the evolution of ever more sophisticated algorithms which are capable of tackling increasingly complex problems. We will now conclude the chapter with an outline of the development of coupled models which include heat-flow. This is not intended to be exhaustive; merely an indication of some of the possibilities.

The melting of ice which was modelled by Butler and Johns [5.2] was probably the earliest example of a coupled model. This and its enhancement by Hurst and Pulko [5.30] are considered from the viewpoint of mass transfer in the next chapter.

(i) ESD prediction

The prediction of electrostatic damage (ESD) sensitivity in integrated circuits was the next coupled model to be tackled, but as it was done as part of a consultancy, it has not been formally published. Integrated circuits can be damaged by very short duration high voltage pulses which have an exponential decay. If such a pulse reaches one of the conductive tracks in an integrated circuit then a depletion can quickly extend between neighbouring tracks. If these touch then the device enters a punch-through regime along a high-field line such as that shown (marked 'ab') in figure 5.11. This electromagnetic response to a rapid pulse was first modelled using a TLM field solver. It was assumed that current flow would be restricted to the region which was in punch-through. The power dissipation data was used as input to a thermal solver which predicted temperature maxima at either point 'a' or point 'b' in figure 5.11 depending on the polarity of the applied pulse. This was completely consistent with actual ESD component

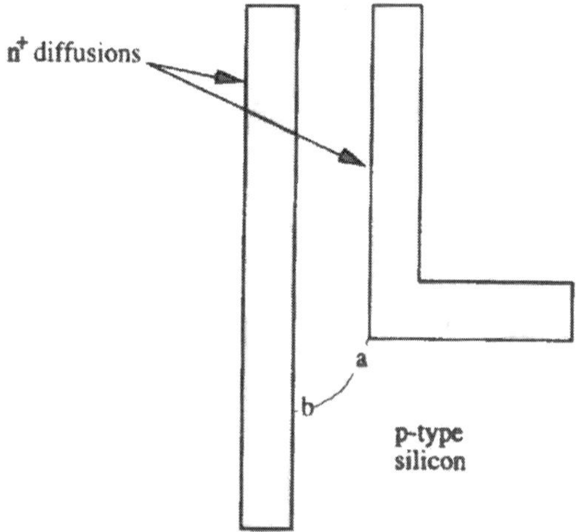

Figure 5.11 n^+ diffusions in an integrated circuit for modelling ESD.

destruction which showed polarity/magnitude dependent burn-holes in the silicon at one or other of these locations.

This work was not taken any further, but it would be quite possible to incorporate a TLM EM/thermal solver in conventional integrated design software as a means of predicting ESD sensitivity.

(ii) The firing of china-ware

The processes which take moulded clay and transform it into china-ware by firing is another example of something that can be treated using a coupled model. This has been done by Pulko and colleagues and involves multiple modes of heat transfer, material transport and phase transitions, some of which are discussed in detail in the next chapter. There has also been some very exciting work on combining thermal and stress models [5.29, 5.31].

(iii) Microwave heating

Microwave heating is another example of the integration of electromagnetic and thermal solvers. This was studied by Desai *et al* [5.32] in relation to microwave cooking and extended to industrial heating by Trenkiç [5.33]. A lossless TLM solver can be used to confirm the resonant frequency of a microwave cavity. The excitation of a field solver at the resonant frequency can be used to identify the regions of maximum microwave power. If a foreign material (such as an item of food) is introduced into a microwave oven, then the different permittivity will affect the field pattern. The dielectric interface leads to impedance mismatch so that there will be a standing field pattern inside the introduced material.

Electromagnetic energy leads to heating through dielectric losses. This appears in the expression for relative permittivity:

$$\varepsilon_r = \varepsilon' - j\varepsilon'' \text{ (where } j = \sqrt{-1}) \tag{5.22}$$

ε' represents a current which is 90° out of phase with the applied voltage and therefore does not lead to power dissipation. On the other hand ε'' represents an in-phase current and does lead to heating. The effect is frequently quoted as loss angle, $\tan\delta = \varepsilon''/\varepsilon'$. The loss angle is a function of frequency, but values which are smaller than 10^{-5} are representative of good (lossless dielectrics) which will not heat in a microwave oven. Values of $\tan\delta$ which are larger than about 10^{-3} are quite definitely lossy and will heat.

The power dissipation due to dielectric loss is then transferred into a TLM thermal solver as a set of current generators. It must be remembered that in the case of microwave cooking, these may be strongly temperature dependent, particularly if the heating process brings about a chemical or physical change.

(iv) Current hogging in semiconductor devices

The TLM solution of thermal effects in semiconductor devices can be very time consuming. The external electrical conditions act as input to an electrical transport model. The local currents and potentials at any time then provide power dissipation data for the thermal solver. This is usually a highly non-linear problem which can be numerically unstable. Russell and Webb [5.34] have used a novel approach involving a circuit analogue of a heterojunction bipolar transistor (HBT) which is analysed using a SPICE simulator. Power dissipation information is passed to a TLM thermal solver. Information from the thermal solver is used in turn to update temperature dependent parameters in the SPICE model. This has successfully simulated the phenomenon of current hogging which can lead to failure in HBTs.

References

[5.1] P.B. Johns, A simple, explicit and unconditionally stable numerical routine for the solution of the diffusion equation, *Int. Jnl for Numerical Methods in Engineering*, 11 (1977) 1307–1328.

[5.2] G. Butler and P.B. Johns, The solution of moving boundary heat problems using the TLM method of numerical analysis, *Numerical Methods in Thermal Problems* (Ed. R.W. Lewis and K. Morgan), Pineridge Press, Swansea, 1979, pp. 189–195.

[5.3] D. de Cogan and S.A. John, The Calculation of Temperature Distribution in Punch-Through Structures During Pulsed Operation Using the Transmission Line Modelling (TLM) Method, *J. Phys. D.*, 15 (1982) 1979–1990.

[5.4] D. de Cogan and S.A. John, A Two Dimensional TLM Model for the Punch-Through Diode , *J. Phys. D.*, 18 (1985) 507–515.

[5.5] D. de Cogan and S.A. John, Failure Modes in Punch-Through Diodes, *J. Phys. D.*, 18 (1985) 497–505.

[5.6] D. de Cogan, A.K. Shah and M. Henini, Variable Mesh TLM Modelling of Heat Flow in Semiconductors, *Proc. of Fourth Int. Conf. on Numerical Analysis of Semiconductor Devices*, (NASECODE IV), Boole Press, Dublin 1985, p. 255.

[5.7] D. de Cogan and A.K. Shah, Stub TLM Modelling of Heat Flow in Semiconductors, *J. Phys. D.*, 19 (1986) 721–715.

[5.8] S.H. Pulko, A. Mallik and P.B. Johns, Application of transmission line modelling (TLM) to thermal diffusion in bodies of complex geometry, *Int. Jnl. for Numerical Methods in Engineering*, 23 (1986) 2303–2312.

[5.9] D. de Cogan, P. Enders and X. Gui, Impedance Transformations and mesh coarsening in TLM heat flow modelling, *Numerical Heat Transfer*, (Part B) 21 (1992) 327–342.

[5.10] S.H. Pulko and P.B. Johns, Modelling of thermal diffusion in three dimensions by the transmission line matrix method and the incorporation of non-linear thermal properties, *Communications in Applied Numerical Methods*, 3 (1987) 571–579.

[5.11] D. de Cogan and M. Henini, TLM modelling of the thermal behaviour of conducting films on insulating substrates, *J. Phys. D: Appl. Phys.*, 20 (1987) 1445–1450.

[5.12] R. Ait-Sadi and P. Naylor, An investigation of the different TLM configurations used in the modelling of diffusion problems, *Int. Jnl of Numerical Modelling*, **6** (1993) 253–268.

[5.13] X. Gui and P.W. Webb, A comparative study of two TLM networks for the modelling of diffusion processes, *Int. Jnl of Numerical Modelling*, **6** (1993) 161–164.

[5.14] S.H. Pulko, I.A. Halleron and C.P. Phizacklea, Substructuring of space and time in TLM diffusion applications, *International Journal of Numerical Modelling*, **3** (1990) 207–214.

[5.15] S.H. Pulko and C.P. Phizacklea, A thermal model for cyclic glass pressing processes using transmission line modelling, *Proc. Inst. Mech. Engineers*, **205** (1991) 187–194.

[5.16] R.M. Witwit, Meshing techniques for TLM diffusion problems, *PhD thesis*, Hull University (UK) 1994.

[5.17] R.M. Witwit, A.J. Wilkinson and S.H. Pulko, A TLM model for anisotropic heat-flow *Proceedings of the Thirteenth International Conference on Applied Informatics*, Innsbruck 1995, pp 87–90.

[5.18] D. de Cogan and M. Henini, TLM Modelling of solder voids in power semiconductors, *IEEE Proc. Components, Hybrids and Manufacturing Technology*, CHMT-10 (1987) 440–445.

[5.19] D. de Cogan and M. Henini, Thermal effects in thin conducting films on silica and alumina substrates under near adiabatic conditions, *J. Phys. D.*, **20**, (1987)1451–1453.

[5.20] S.V.G. Vardigans, D. de Cogan and M. Henini, A transient thermal failure mode in metal film resistors, *J. Phys. D:* **20** (1987) 1454-1456.

[5.21] D. de Cogan, Non-linear thermal effects in metal/non-metal systems, *Int. Jnl. of Electronics*, **77** (1994) 459-456.

[5.22] E.W. Williams, H.C. Patel, D. de Cogan and S.H. Pulko, TLM modelling of thermal processes in magneto-optic multi-layered media, *J.Phys. D: Appl. Phys.*, **29** (1996) 1124–1132.

[5.23] T. Phan, S. Dilhaire, V. Quintard, D. Lewis, W. Claeys and J.C. Batsale, Modelling and measurement of micrometric interconnect's transient temperature. Application to thermal condctivity identification., *Proc. 2nd THERMINIC Workshop, Budapest*, Sept. 1996, pp 153–159.

[5.24] S.H. Pulko, A. Mallik, R. Allen and P.B. Johns, Automatic time-stepping in TLM routines for the modelling of thermal diffusion processes, *Int. Jnl. of Numerical Modelling*, **3** (1990) 127–136.

[5.25] P.W. Webb and X. Gui, Time-step changes in TLM diffusion modelling, *Int. Jnl. of Numerical Modelling*, **5** (1992) 251–257.

[5.26] I.A.D. Russell and P.W. Webb, Automatic time-step control in TLM diffusion modelling for problems with time-varying load conditions, *Int. Jnl. of Numerical Modelling*, **9** (1996) 417–428.

[5.27] S.H. Pulko, W.A. Green and P.B. Johns, An extension of the application of transmission line modelling (TLM) to thermal diffusion to include non-infinite heat sources, *Int. Jnl. for Numerical Methods in Engineering*, **24** (1987) 1333–1341.

[5.28] D. de Cogan, Thermal radiative boundaries in TLM, *Int. Jnl. of Numerical Modelling*, **6** (1993) 165–166.

[5.29] P. Langley, S.H. Pulko and A.J. Wilkinson, A TLM model of transient two-dimensional stress propagation, *Int. Jnl. of Numerical Modelling*, **9** (1996) 429–443.

[5.30] A.I. Hurst and S.H. Pulko, TLM treatments of changes of phase, *International Journal of Numerical Modelling*, **7** (1994) 201–207.

[5.31] H. Newton, TLM models of deformation and their application to vitreous china ware during firing, *PhD thesis*, Hull University 1994.

[5.32] R.A. Desai, C. Christopoulos, P. Naylor, J.M.V. Blanshard and K. Gregson, Computer modelling of microwave cooking using a transmission line model, *IEE Proceedings A*, **139** (1992) 30–38.

[5.33] V. Trenkiç, Numerical modelling of industrial microwave heating using the transmission line method, *BSc thesis*, Universities of Nottingham (UK), Nis (Yugoslavia) 1992.

[5.34] I.A.D. Russell and P.W. Webb, Simulation of electro-thermal instability in microwave devices using the TLM method in "TLM: The wider applications" (Ed. D. de Cogan), School of Information Systems (UEA) 1996 pp. 4.1–4.10.

Chapter 6

The application of TLM algorithms to particle diffusion

Introduction

To a first approximation the equations for heat-flow and matter diffusion are equivalent and it was obvious to Johns that the concepts which were developed in his paper [6.1] could be applied to both, provided that the distributed product $(R_d C_d)^{-1}$ was set equal to the diffusion constant. There is some evidence that he presented details of the application to food engineering in 1984, but the first serious work was undertaken by M. Hendrickx as part of his PhD research on the absorption of water by white rice at the University of Louvain in Belgium [6.2–6.5]. The state of knowledge in the area of matter diffusion as at 1987 was then summarised by Johns and Pulko [6.6]. TLM has now been applied in other areas involving particle diffusion. Several of these are discussed here, although the chromatographic process is deferred to the next chapter since it combines diffusion with drift.

Theory

The requirement to have $D = (R_d C_d)^{-1}$ seems to create an ambiguity, since TLM normally requires an independent definition of the nodal parameters R and $Z = \Delta t / C$. The normal approach to algorithm development has been to decide first on the nodal dimensions, which in turn depends on the density of information that is required from a simulation.

Therefore, we can write

$$\frac{\Delta x^2}{D} = \tau \tag{6.1}$$

This gives us a pointer for the choice of the discretisation time, which we normally arrange to be $\Delta t \leq \tau$.

If we choose to operate with unit nodal capacitance then $Z = \Delta t$.

$$\text{in which case } C_d = \frac{1}{\Delta x} \text{ and therefore } \frac{1}{D} = \frac{R}{\Delta x}\frac{1}{\Delta x} \tag{6.2}$$

so that the nodal resistance $R = \Delta x^2/D = \tau$

In a one-dimensional TLM algorithm the reflection coefficient is $\rho = R/(R+Z)$.

Thus the choice $\Delta t = \tau$ corresponds to $\rho = 0.5$ (the limit when explicit finite difference and TLM are identical).

In spite of this, the $D = (R_dC_d)^{-1}$ ambiguity can be worrisome.

Let us imagine that we had a material with a diffusion constant of $10^{-2}\,\text{cm}^2\ \text{s}^{-1}$ and let us also suppose that we choose $\Delta x = 1\text{cm}$.

If $C = 1$ then $C_d = 1$. Therefore $R_d = R = 100$

If on the other hand $C = 10$ then $R = 10$

We could then use either of these conditions to model a problem where material diffused through a medium which separated a constant source of diffusant at one side and a constant sink at the other. In both of these simulations we will obtain the same transient behaviour and the same spatial distribution at $t = \infty$. If however, we treat the equilibrium case as an Ohm's law problem then it is obvious that the rate of material flow when $R = 10$ will be ten times the value it would have when $R = 100$. This suggests that something is not quite right.

We could attempt to resolve this by approaching the problem from the steady-state:

If we have a one-dimensional diffusion medium with a concentration difference ΔC across its length, L then we can write the steady state flux as:

$$\text{particles/sec/area} = D\frac{\Delta C}{L}$$

If we use the analogues: flux $\equiv J$ (current density) and $\Delta C \equiv V$ then:

$$\text{Current} \equiv AD\,\frac{\Delta C}{L} = \frac{V}{R_{\text{total}}}$$

In this case of unit area $R_{\text{total}} = L/D$ or $R_d = 1/D$ and therefore $C_d = 1$ which indicates that those who have adopted $C_d = 1$ were in fact physically consistent.

Diffusion with chemical reaction

de Cogan and Henini [6.7] applied a TLM diffusion algorithm to model periodic precipitation phenomena. This represents a moving boundary problem with chemical reaction. In 1896 Liesegang [6.8] observed the phenomenon of periodic precipitation. He noted that when a drop of concentrated silver nitrate is allowed to diffuse into a gel containing potassium dichromate the resulting precipitate was not uniform but proceeded by a series of concentric rings. These were separated by clear zones. It was found that the distance between successive precipitates was related by:

$$\frac{x_{n+1}}{x_n} = C_1 (\text{a constant}) \tag{6.3}$$

The distance to the n-th precipitate and the time elapsed before its formation was related by

$$\frac{x}{t_n^{1/2}} = C_2 \ (\text{a constant}) \tag{6.4}$$

There are many theories in chemistry to account for these observations [6.9–6.12]. These generally depend on the law of mass action adapted for a sparingly soluble material:

Let us start by considering a chemical equilibrium:

$$aA + bB \rightleftharpoons cC + dD$$

(where A, B, C and D are the species and a, b, c and d are the ratios in which they react).

Under normal circumstances the ratio of forward and reverse velocities of a chemical reaction can be expressed by

$$K_{\text{eq}} = \frac{[C]^c[D]^d}{[A]^a[B]^b} \tag{6.5}$$

the square brackets indicate concentrations.

If a material, AB is sparingly soluble in water then there is an equilibrium between the solid AB (precipitate) and the ions A^+ and B^- in solution

$$AB \rightleftharpoons A^+ + B^- \tag{6.6}$$

Since the equilibrium is due to a small fraction, α of AB dissociating, then we can write the equilibrium as

$$K_{eq} = \frac{\alpha(\text{of } A^+)\,\alpha(\text{of } B^-)}{(1-\alpha)\,(\text{of } AB)} \tag{6.7}$$

Now if $\alpha \ll 1$ then the concentration of AB is almost unchanged. The equilibrium is then determined by $[A^+][B^-]$ which is called the ***solubility product*** and is a widely tabulated parameter for semi-soluble materials.

In the case of silver dichromate (the precipitate in the Liesegang experiment) the equilibrium is:

$$Ag_2Cr_2O_7 \rightleftharpoons 2Ag^+ + Cr_2O_7^{2-} \tag{6.8}$$

and the solubility product is $[Ag^+]^2\,[Cr_2O_7{}^{2-}]$ and has a value of 2×10^{-7} de Cogan and Henini used diffusion coefficients of 1.6×10^{-5} and 1×10^{-5}cm s^{-1} for $AgNO_3$ and $K_2Cr_2O_7$ respectively. Using $\Delta x = 1$mm and t = 10s in a one-dimensional TLM routine they allowed the silver nitrate to diffuse into the jelly. Whenever the concentrations at a node were such that $[Ag^+]^2\,[Cr_2O_7{}^{2-}]$ was greater than or equal to 2×10^{-7} then the presence of a precipitate was flagged and the species currently at that node were removed. This meant that fresh reactants had to diffuse in from adjacent nodes. The application of this simple model for a range of $AgNO_3$ concentrations confirmed the validity of eqns (6.3, 6.4).

It is worth noting that the $t_n^{1/2}$ rule is common in moving boundary phenomena which could be easily modelled using TLM. These include the freezing of water on lakes and the oxidation of silicon.

Modelling of phase transitions

It is not clear whether the subject of phase transitions should come under heat-flow or particle diffusion. However, as they have close relationships with the preceding sections and since they frequently involve moving boundaries, this important topic is included here.

The first known example of the modelling of a phase transition using TLM was the paper by Johns and Butler [6.13] when they simulated the melting of ice. This process involves the absorption of latent heat to affect a change from the solid to liquid state. After the transition, the interface moves and heat continues to be absorbed into the remaining ice. The demonstration cited in the paper involved a piece of ice in the shape of a letter 'A'. It was shown that where heat could enter the ice by more than one direction (e.g. in the vicinity of a corner) then the melting process occurred more rapidly. This has much in common with the diffusive degradation of images which is discussed in Chapter 10.

The other known TLM treatment of phase change has been by Hurst and Pulko [6.14] as part of their study of the firing of clay in the manufacture of china-ware. In their work they have generalised the treatment and have removed some of the restrictions that were inherent in the original approach. Specifically, they addressed exothermic processes, where a phase-change leads to the generation of heat. They also allowed the temperature of the node undergoing transformation to vary and to contribute to the change of phase.

A phase transition either absorbs or emits heat, but this occurs at a specific rate, which is determined by a rate constant, K. This depends on temperature according to the Arhennius equation:

$$K = Ae^{-B/T} \text{ (where } T \text{ is in Kelvin)} \tag{6.9}$$

A and B are material constants which in the case of firing clay depend on grain-size and crystal structure.

In a three-dimensional TLM treatment the temperature at a node is given by:

$$T(x) = \frac{\displaystyle\sum_{j=1}^{6} \frac{2_k i V_j(x)}{R_j + Z_j} + {}_k I(x)}{\displaystyle\sum_{j=1}^{6} \frac{1}{R_j + Z_j}} \tag{6.10}$$

The current generator, ${}^k I(x)$ is given by:
K times the total remaining latent heat available at iteration k *for exothermic* or K times the total remaining latent heat required at iteration k *for endothermic*. Thc net heat in a node is ${}^i V(x) - {}^r V(x)/Z$.

This accumulates iteration by iteration until sufficient is present for the latent heat requirement of the nodal volume *and* the heat associated with any change in specific heat of the node. Heat not required to effect a change of state is allowed to change the temperature of the node (something that was not considered in the original Johns/Butler model. This of course gives rise to a change in the rate constant, K.

If external conditions cause the influx of heat into a node which is undergoing an endothermic change to decline then the requirement is drawn from the node itself. The nodal temperature drops and with it the rate constant, K.

TLM modelling of diffusion effects in thin films for VLSI applications

In 1993 Gui *et al* published a paper on the effects of grain boundary and interfacial diffusion effects in thin films [6.15]. These have particular significance in very large-scale integrated circuit (VLSI) technology, since diffusion along the grain boundaries of the polycrystalline material is more rapid than in the equivalent bulk. This research has developed to consider electromigration, impurity diffusion in poly-silicon and in silicides as well as the effectiveness of diffusion barrier layers. The initial paper [6.15] developed a three-dimensional TLM model with a generalised multi-grid and a time-step control algorithm as outlined in [6.16].

Figure 6.1 shows the system which has been modelled. There are grains of thickness, h which are surrounded by boundaries of width $\delta/2$. The grains, boundaries, interface and substrate have their respective diffusion constants, D_g, D_b, D_i and D_s. Because of the symmetry of the problem it is possible to model the system by considering only the region of length, L which is shaded. This is shown in figure 6.2 along with suggested boundaries which would be used in a TLM simulation.

The constant value source is equivalent to a constant voltage source in TLM and is normalised to unity. The basis of this and several other boundary definitions are discussed in other papers which are reviewed below. The analysis also used a normalised (dimensionless) time $T = D_g\Delta t/L^2$.

Several simulations were performed. In all of these the following realistic conditions were used: $L/h = 1$, $d/h = 5 \times 10^{-3}$, $D_b/D_g = 10^4$ and $D_s/D_g = 10$.

When the ratio D_i/D_g is small ($i.e. = 1$) then the problem can be considered as a set of isolated grains. When it is 100 times larger this is equivalent to the film/substrate interface acting as a fast diffusion path. When the interface and substrate are impenetrable to atomic diffusion, then the rear of the film effectively becomes a diffusion barrier.

These concepts have also been extended to grain-boundary diffusion in thin films with micro-structural details [6.17]. This paper pays much attention to the boundary descriptions that play a fundamental role in grain

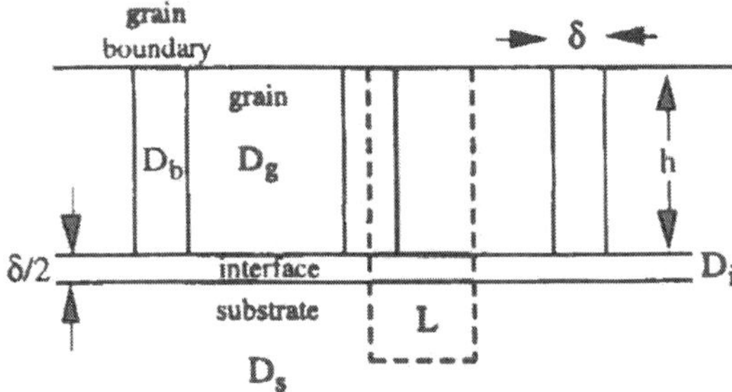

Figure 6.1 The film on substrate model with interface and grain-boundaries as modelled by Gui *et al* [6.15].

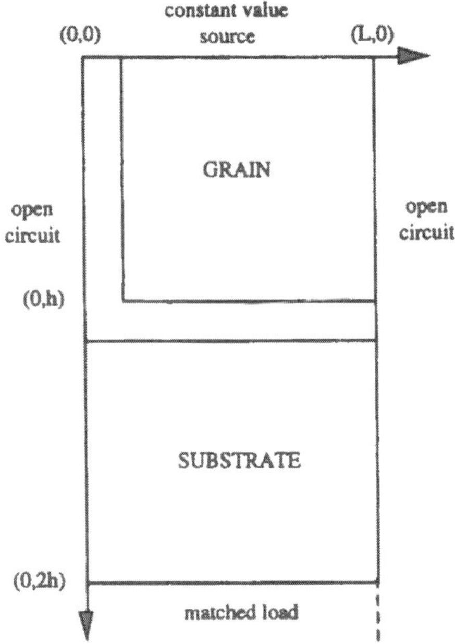

Figure 6.2 The reduced grain-boundary system which was modelled using TLM.

boundary modelling. They have considered a range of microstructures from simple path to meander path, to single-branch and multi-branch interfaces which might constitute a diffusion barrier in titanium nitride on silicon. The impurity penetration is shown to decrease as the diffusion path complexity increases. These simulations have been implemented for a more realistic microstructure representation generated by GROFILMS, a Monte Carlo film growth simulator.

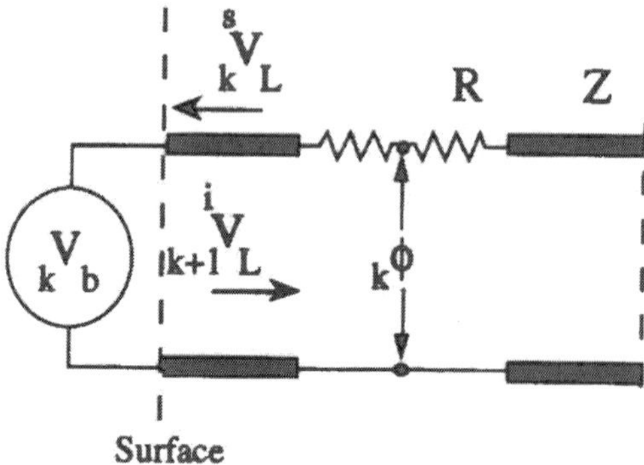

Figure 6.3 The TLM representation of a Robins boundary.

In subsequent publications the various boundary conditions which were initially assumed are discussed in detail [6.18]. The Robin boundary is a mixed boundary condition:

$$D\frac{\partial C(L,t)}{\partial x} + \alpha\, C(L,t) = 0 \quad \text{for } t \geq 0 \tag{6.11}$$

and has been subject of special consideration. In this case α is a parameter characterising the out-diffusion mobility from the region of interest. This condition can be interpreted as:

(1) the difference between concentrations in the problem space and outside driving matter flux through the surface,
(2) analogous to Ohm's law,
(3) a continuity equation comprising diffusion flux and convection flux.

A TLM representation of the boundary is shown in figure 6.3 and it can be seen that the condition can be expressed in finite difference form as:

$$D\frac{{}_kV_b - {}_k\phi}{\Delta x/2} + \alpha_k V_b = 0 \tag{6.12}$$

The magnitude of the external voltage source can be determined from this equation as:

$${}_kV_b = \frac{2D}{2D+\alpha\Delta x}\,{}_k\phi \tag{6.13}$$

This expression can then be used to calculate the pulse incident from the surface at the next time-step:

$$ {}_{k+1}^{\;\;i}V_L(1) = {}_kV_b - {}_k^rV_L(1) \tag{6.14} $$

The validity of this approach has been demonstrated using a range of examples. Additionally, the authors have shown how the matched load condition can also be interpreted as a special case of this boundary condition, namely ${}_kV_b = {}_k^rV_L(1)$.

The interpretation of the matched load condition was taken a stage further [6.19]. The authors have followed the derivation of Al-Zeben *et al* [6.20] and expressed the flux as:

$$ D\frac{\partial \phi}{dx} = -\frac{IZ\Delta x}{\Delta t} \tag{6.15} $$

The equation for the boundary condition can then be expressed as:

$$ -\frac{IZ\Delta x}{\Delta t} + \alpha V_b = 0 \tag{6.16} $$

This suggests that we can define an equivalent boundary resistance:

$$ R_b = \frac{V_b}{I} = \frac{Z\Delta x}{\alpha\Delta t} \tag{6.17} $$

We can now define the incident pulse at the new iteration as:

$$ {}_{k+1}^{\;\;i}V_L(1) = \rho\,{}_k^rV_L(1) = \left[\frac{R_b - Z}{R_b + Z}\right] {}_k^rV_L(1) = \left[\frac{\Delta x - \alpha\Delta t}{\Delta x - \alpha\Delta t}\right] {}_k^rV_L(1) \tag{6.18} $$

Thus, the boundary voltage at the k-th iteration can be given as:

$$ {}_kV_b = {}_{k+1}^{\;\;i}V_L(1) + {}_k^rV_L(1) = \left[\frac{2\Delta t}{\Delta t - \alpha\Delta t}\right] {}_k^rV_L(1) \tag{6.19} $$

These two equations are sufficient for treating the Robin boundary condition.

Important cases include:

$\alpha \to \infty$	$R_b \to 0$	$\rho = -1$	a short-circuit boundary
$\alpha \to 0$	$R_b \to \infty$	$\rho = 1$	an open-circuit boundary
$\alpha = \dfrac{\Delta x}{\Delta t}$	$R_b = Z$	$\rho = 0$	a matched-load boundary

This last condition means that the diffusing species can move an elemental distance away from the boundary within an elemental time.

The general concepts of grain boundary diffusion have been extended into the area of drift diffusion to model electromigration and thin-film interconnect failure due to high current pulses [6.21, 6.22].

Anisotropic diffusion

Until now there have been almost no examples of using TLM to model real diffusion processes at an atomic level such as the doping of silicon by phosphorous. Diffusion processes of this type are heavily influenced by the crystal structure and proceed by vacancy and/or interstitial mechanisms. The diffusion constant can also be strongly dependent on the local concentration of diffusant. In this case it would be more appropriate to use the anisotropic form of the diffusion equation:

$$\frac{\partial C}{\partial t} = \frac{\partial}{\partial x}\left[D\frac{\partial C}{\partial x}\right] \tag{6.20}$$

This can be expressed as

$$\frac{\partial C}{\partial t} = D(x)\frac{\partial^2 C}{\partial x^2} + \frac{\partial D(x)}{\partial x}\frac{\partial C}{\partial x} \tag{6.21}$$

$$= D(x)\frac{\partial^2 C}{\partial x^2} + \frac{\partial D(x)}{\partial C}\left[\frac{\partial C}{\partial x}\right]^2 \tag{6.22}$$

or as

$$\frac{\partial^2 C}{\partial x^2} = \frac{1}{D(x)}\frac{\partial C}{\partial t} - \frac{1}{D(x)}\frac{\partial D(x)}{\partial C}\left[\frac{\partial C}{\partial x}\right]^2 \tag{6.23}$$

This can be approximated by the TLM expression:

$$\frac{\partial^2 C}{\partial x^2} = R_d C_d \frac{\partial C}{\partial t} + G \tag{6.24}$$

where $G = \frac{1}{D(x)}\frac{\partial D(x)}{\partial C}\left[\frac{\partial C}{\partial x}\right]^2$ is an equivalent source

References

[6.1] P.B. Johns, A simple, explicit and unconditionally stable routine for the solution of the diffusion equation, *Int. Jnl. Numerical Methods in Engineering*, 11 (1977), 1307–1328.

[6.2] C. Engels, M. Hendrikx, M. de Samblanx, I. de Gryze and P. Tobback, Modelling water diffusion during long-grain rice soaking, *J. Food Engineering*, 5 (1985), 55–73.

[6.3] M. Hendrickx, C. Engels, P. Tobback and P. Johns, Transmission line modelling (TLM) of water diffusion in white rice, *J. Food Engineering*, 5 (1986), 269–285.

[6.4] M. Hendrickx, C. Engels and P. Tobback, Three dimensional TLM models for water diffusion in white rice, *J. Food Engineering*, 6 (1987), 187–197.

[6.5] M. Hendrickx, The application of Transmission Line Matrix (TLM) to food engineering PhD thesis, Katholiek Universiteit te Leuven, 1988.

[6.6] P.B. Johns and S.H. Pulko, Modelling of heat and mass transfer in foodstuffs in 'Food structure and behaviour' (Eds. J.M.V. Blanshard and P. Lilliford) Academic Press 1987, pp. 199–218.

[6.7] D. de Cogan and M. Henini, Transmission Line Modelling of the Liesegang Phenomenon, *J. Chem. Soc.*, Faraday Transactions II 83 (1987) 837–841.

[6.8] R.E. Liesegang, *Naturwiss. Wochenschr* 11 (1896) 53.

[6.9] W. Ostwald, Lehrbuch der Algemeinen Chemie, Engelman, Leipzig 1897, p. 778.

[6.10] N.R. Dhar and A.C. Chatterji, *Kolloid Zeit.*, 37 (1925) 2.

[6.11] C. Wagner, *J. Colloid Science*, 5 (1950), 85.

[6.12] S. Prager, *J. Chem. Phys.*, 25 (1956), 279.

[6.13] P.B. Johns and G. Butler, The solution of moving-boundary heat problems using the TLM method of Numerical analysis. *Proc. First International Conference on Numerical Methods in Thermal Problems*, (Eds. R.W. Lewis and K. Morgan), Pineridge Press, Swansea 1979 pp. 189–195.

[6.14] A.I. Hurst and S.H. Pulko, TLM treatments of changes of phase, *International Journal of Numerical Modelling*, 7 (1994), 201–207.

[6.15] X. Gui, S.K. Dew, M.J. Brett and D. de Cogan,Transmission Line Matrix modelling of grain boundary diffusion in thin films, *Journal of Applied Physics*, 74 (1993), 7173–7180.

[6.16] P.W. Webb and X. Gui, Implementation of time-step changes in transmission line matrix diffusion modelling, *International Journal of Numerical Modelling*, 5 (1992), 251–257.

[6.17] X. Gui, L.J. Friedrich, S.K. Dew, M.J. Brett and T. Smy, TLM modelling of grain-boundary diffusion in thin films with microstructural details *Proc. First International Workshop on Transmission Line matrix (TLM) Modelling Theory and Applications*, Victoria BC, August 1995.

[6.18] X. Gui, S.K. Dew and M.J. Brett, TLM treatment of a general diffusion flux boundary condition, *International Journal of Numerical Modelling*, 9 (1996), 327–333.

[6.19] X. Gui, D. de Cogan, S.K. Dew and M.J. Brett, A general expression of boundary conditions in TLM diffusion modelling (to appear in *International Journal of Numerical Modelling*, Vol 10).

[6.20] M.Y. Al Zeben, A.H.M. Saleh and M.A. Al-Omar, TLM modelling of diffusion, drift and recombination of charge carriers in semiconductors, *International Journal of Numerical Modelling*, 5 (1992), 219–225.

[6.21] X. Gui, S.K. Dew and M.J. Brett, Numerical solution of the electromigration boundary value problem under pulsed *dc* conditions (to appear in *Journal of Applied Physics*).

[6.22] X. Gui, S.K. Dew and M.J. Brett, Thermal simulation of thin film interconnect failure caused by high current pulses (to appear in *IEEE Electron Device Letters*).

Chapter 7

Diffusion/drift models

Introduction

In chapter 5 we looked at TLM models of forced convection, being the mode by which heat is transferred between a fluid and a thermally conductive solid. This chapter looks at TLM models of forced convection from the point of view of the thermal processes within the fluid. There are many physical problems where a diffusion process is accompanied by a drift or overall movement of the diffusing species in one direction with respect to some fixed external reference. This chapter starts by outlining the basics of diffusion/drift theory. We then describe some of the techniques that have been used in TLM implementations and conclude with a discussion on the stability and accuracy of diffusion/drift algorithms.

Background theory

The convection, advection or drift-diffusion equation in one dimension is given by:

$$\frac{\partial C(x,t)}{\partial t} = -\frac{\partial (vC(x,t))}{\partial x} + \frac{\partial}{\partial x}\left[D\frac{\partial C(x,t)}{\partial x}\right] \tag{7.1}$$

which if the drift velocity, v and diffusion constant, D are invariant becomes:

$$\frac{\partial C(x,t)}{\partial t} = -v\,\frac{\partial (C(x,t))}{\partial x} + D\,\frac{\partial^2 C(x,t)}{\partial x^2} \tag{7.2}$$

These equations are the drift modified forms of the classic Fourier-Fick equations and using an approach due to Enders [7.1] will be shown to arise because of a non-local dependence of transition probabilities. The situation with locally varying transition probabilities leads to the *forward Kolmogorov or Fokker-Planck* equation:

$$\frac{\partial C(x,t)}{\partial t} = -\frac{\partial (vC(x,t))}{\partial x} + \frac{\partial^2 [DC(x,t)]}{\partial x^2} \tag{7.3}$$

At the discrete event level we can develop a one-dimensional description that involves transition probabilities f, b and s corresponding to forward, backward or stay so that the concentration of node (x) at time $k+1$ is given by:

$${}_{k+1}C(x) = f_k C(x-1) + b_k C(x+1) + s_k C(x) \tag{7.4}$$

The first few values of the fundamental solution of this equation are shown in Table 7.1.
We can also consider Fick's first law in a most general inhomogeneous form extended by drift/convection:

$$J(x,t) = -D(x)\,\frac{\partial C(x,t)}{\partial x} + v(x)\,p(x)\,C(x,t) \tag{7.5}$$

The discrete analogue is derived by taking the current density from node $x-1$ to its right-hand cell boundary of an auxiliary concentration C_b. Thus

$${}_{k+}J(x-) = -\frac{2D(x-1)}{\Delta x(x-1)}[C_b - {}_kC(x-1)] + V(x-1)p(x-1)\,{}_kC(x-1) \tag{7.6}$$

and equating it to that from the cell boundary to node (x)

Table 7.1

k	$x-3$	$x-2$	$x-1$	x	$x+1$	$x+2$	$x+3$
0				1			
1			b	s	f		
2		b^2	$2sb$	$2bf+s^2$	$2sf$	f^2	
3	b^3	$3sb^2$	$3b^2f+3s^2b$	s^3+6sbf	$3bf^2+3s^2f$	$3sf^22$	f^3

$$_{k+}J(x-) = -\frac{2D(x)}{\Delta x(x)}[_kC(x) - C_b)] + V(x)\,p(x)\,_kC(x) \qquad (7.7)$$

C_b can then be eliminated to give

$$_{k+}J(x-) = -\left[\frac{(\nu(x-1)\nu(x)}{\nu(x-1)+\nu(x)}\right]((_kC(x) - {_kC}(x-1)) + \frac{\nu(x)v(x-1)p(x-1)_kC(x-1) + \nu(x-1)v(x)p(x)_kC(x)}{\nu(x-1)+\nu(x)} \qquad (7.8)$$

where $\nu(x) = 2D(x)/\Delta x(x)$

The continuity equation, which can be more easily discretised in integral form yields:

$$p(x)\Delta x(x)\,\frac{_{k+1}C(x) - {_k}C(x)}{\Delta t} + {_{k+}}J(x+) - {_{k+}}J(x-) = -p(x)\Delta x(x)A(x)_kC(x) \qquad (7.9)$$

Inserting this back into eqn (7.7) and its counterpart for $_{k+}J(x+)$ we obtain:

$$p(x)\,\frac{\Delta x(x)}{\Delta t}\,_{k+1}C(x) = \alpha_1 + \alpha_2 + \alpha_3 \qquad (7.10)$$

where

$$\alpha_1 = \frac{\nu(x)}{\nu(x)+\nu(x+1)}[\nu(x+1) - v(x+1)p(x+1)]_kC(x+1)$$

$$\alpha_2 = \frac{\nu(x)}{\nu(x)+\nu(x-1)}[\nu(x-1) + v(x-1)p(x-1)]_kC(x-1)$$

$$\alpha_3 = \left\{\Delta x(x)p(x)\left[\frac{1}{\Delta t} - A(x)\right] - \frac{\nu(x-1)}{\nu(x)+\nu(x-1)} \times [\nu(x) - v(x)p(x)] - \frac{\nu(x+1)}{\nu(x)+\nu(x+1)}[\nu(x) + v(x)p(x)]\right\}_kC(x)$$

Eqn (7.10) can be compared term-by-term with eqn (7.4). Some interesting points become clear.

- the transition probabilities f, b and s are certainly non-local
- drift is due to an asymmetric shift in f and b
- there is a probability to stay in the same cell even if $A(x) = 0$

The intermediate state between this and a full TLM implementation starts with a correlated random walk:

$$\begin{aligned}
{}_{k+1}^{i}V_L(x) &= \tau({}_k^iV_L(x-1) + {}_k^iV_L(x+1) + (\rho^2-\tau^2)\,{}_{k-1}^{i}V_L(x) \\
{}_{k+1}^{i}V_R(x) &= \tau({}_k^iV_R(x-1) + {}_k^iV_R(x+1) + (\rho^2-\tau^2)\,{}_{k-1}^{i}V_R(x) \qquad (7.11)\\
{}_{k+1}^{i}\phi_L(x) &= \tau\left[{}_k^i\phi_L(x-1) + {}_k^i\phi_L(x+1)\right] + (\rho^2-\tau^2)\,{}_{k-1}^{i}\phi_L(x)
\end{aligned}$$

The normal implementation breaks symmetry by having the reflection and transmission coefficients different for the left and right-moving pulses [7.3].

$$\begin{aligned}
{}_{k+1}^{i}V_L(x) &= \rho_L\,{}_k^iV_L(x-1) + \tau_R\,{}_k^iV_L(x-1) \\
{}_{k+1}^{i}V_R(x) &= \rho_R\,{}_k^iV_R(x+1) + \tau_L\,{}_k^iV_R(x+1)
\end{aligned} \qquad (7.12)$$

We apply the condition that $\rho_L + \tau_L = \rho_R + \tau_R = 1$ and interpret these coefficients in terms of symmetric coefficients as:

$$\rho_L = \rho - \alpha \quad \tau_L = \tau + \alpha \quad \rho_R = \rho + \alpha \quad \tau_R = \tau - \alpha$$

The expressions in eqn (7.11) can be replaced by the appropriate asymmetrical ones. Because of similarities only the nodal potential is given below, but in two different forms:

$$\begin{aligned}
{}_{k+1}\phi(x) = \tau\left[({}_k\phi(x-1) + {}_k\phi(x+1)\right] + (\rho^2-\tau^2)\,{}_{k-1}\phi(x) \\
+\alpha\left[{}_k\phi(x-1) - {}_k\phi(x+1)\right]
\end{aligned} \qquad (7.13a)$$

$${}_{k+1}\phi(x) = \tau_L\,{}_k\phi(x-1) + \tau_R\,{}_k\phi(x+1) + (\rho_R\rho_L - \tau_R\tau_L)_{k-1}\phi(x) \qquad (7.13b)$$

These two expressions have correlation coefficients:
$(\rho^2-\tau^2)_{k-1}\,\phi(x)$ in eqn (7.13a)
$(\rho_R\rho_L - \tau_R\tau_L)_{k-1}\phi(x)$ in eqn (7.13b)
It is particularly interesting to note that these are both independent of the bias, α.

It will also be noted that the effect of scattering asymmetry is to add a term to the standard correlated random walk which is a function of the difference in the potentials of adjacent nodes.

If we define $\rho_L = \Delta x S_L$ and $\rho_R = \Delta x S_R$ then in the continuum limit, where $\Delta x = c\Delta t \to 0$ (where c is a constant) eqn (7.13a) becomes:

$$\frac{\partial^2\phi}{\partial t^2} + c(S_R + S_L)\frac{\partial\phi}{\partial t} = c^2(S_R - S_L)\frac{\partial\phi}{\partial x} + c^2\frac{\partial^2\phi}{\partial x^2} \qquad (7.14)$$

Thus the phenomenological parameters S_R and S_L are associated with the diffusivity and the drift velocity.

$$S_R = \frac{c+v}{2D} \quad \text{and} \quad S_L = \frac{c-v}{2D} \tag{7.15}$$

If we do not use the continuum limit, then the Taylor expansion of eqn (7.13a) yields:

$$\left[\frac{\tau\Delta t}{2\rho}\right]\frac{\partial^2\phi}{\partial t^2} + \frac{\partial\phi}{\partial t} = \left[\frac{\alpha\Delta x}{\rho\Delta t}\right]\frac{\partial\phi}{\partial x} + \left[\frac{\tau\Delta x^2}{2\rho\Delta t}\right]\frac{\partial^2\phi}{\partial x^2} \tag{7.16}$$

In this case $\alpha\Delta x/\rho\Delta t \equiv v$, which provides a TLM-like definition for α.

$$\alpha = \rho\frac{v}{c} = \left[\frac{R}{R+Z}\right]\left\{\frac{\text{drift velocity}}{\text{propagation velocity}}\right\} \tag{7.17}$$

Practical implementations and applications

The earliest approaches to a TLM drift/diffusion treatment [7.4, 7.5] involved a step-wise movement of the flowing species. It assumed that the two processes (diffusion and drift) could be treated independently. This was based on a restriction to cases where the value of $1/\sqrt{D\Delta t}$ was very significantly different from the flow rate.

If Δx and Δt are the space and time discretisations for diffusion then we can express the flow-rate as:

$$U = \frac{\Delta x}{m\Delta t} \tag{7.18}$$

Thus, we leave the diffusion process to operate on the existing mesh After m time-steps have elapsed we take the contents of each node and move them along by one mesh-point in the direction of flow; during the critical moment between time-steps $m-1$ and m the value of node $x-1$ is transferred to node x (assuming that flow is in the x direction). Thus ${}_{m-1}C(x-1) \rightarrow_m C(x)$.

This approach did not prove to be particularly successful as can be seen in the results of de Cogan and Henini [7.5]. It was quickly superseded by the network approach due to Saleh [7.6].

Figure 7.1 shows the basic TLM node that simulates mass diffusion and flow. It is merely the normal diffusive node with the addition of a current generator. The node is shown in terms of its lumped parameters in figure 7.2. The voltage across the inductor and resistor gives

$$-\frac{\partial V}{\partial x}\frac{\Delta x}{2} = RI_x + L_d\frac{\Delta x}{2}\frac{\partial I_x}{\partial t} \tag{7.19}$$

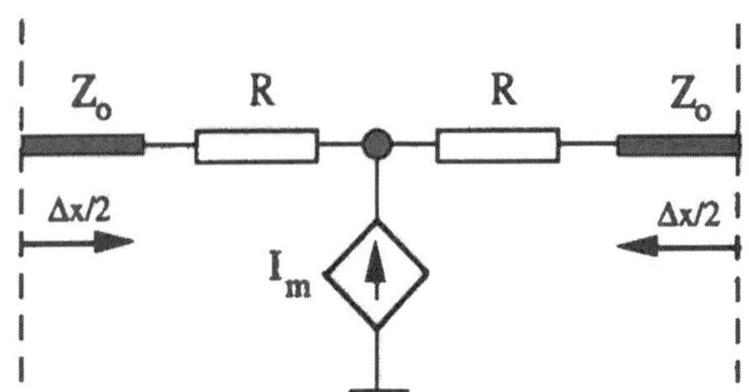

Figure 7.1 a TLM drift-diffusion node.

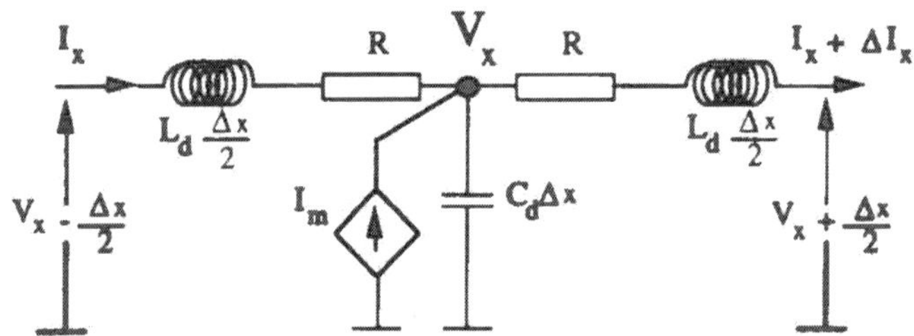

Figure 7.2 The lumped equivalent circuit for the TLM drift-diffusion node.

or with $R_d = 2R/\Delta x$

$$-\frac{\partial V}{\partial x} = R_d I_x + L_d \frac{\partial I_x}{\partial t} \tag{7.20}$$

Current analysis gives

$$\frac{\partial I}{\partial x} \Delta x = I_m + C_d Dx \frac{\partial V_x}{\partial t} \tag{7.21}$$

by taking the current generator $I_m = g_m \Delta V_x$ where $\Delta V_x = (V(x+1) - V(x-1))/2$ one gets:

$$-\frac{\partial I_x}{\partial x} = C_d Dx \frac{\partial V_x}{\partial t} - g_m \frac{\partial V_x}{\partial x} \tag{7.22}$$

The usual argument about neglecting the inductive term is applied and the combination of these two equations then gives:

$$\frac{\partial^2 V}{\partial x^2} + g_m R_d \frac{\partial V}{\partial x} = R_d C_d \frac{\partial V}{\partial t} \tag{7.23}$$

This can be compared with a conventional diffusion/drift equation such as that used in semiconductor transport

$$\frac{\partial^2 C(x,t)}{\partial x^2} + \frac{U}{D}\frac{\partial C(x,t)}{\partial x} = \frac{1}{D}\frac{\partial C(x,t)}{\partial t} \tag{7.24}$$

where D is the diffusion constant and U is the drift velocity. The comparison yields the following analogies:

$$D \equiv \frac{1}{R_d C_d} \qquad U \equiv \frac{g_m}{C_d}$$

$$C(x,t) \equiv V \qquad \frac{\partial}{\partial x} C(x,t) \equiv -I R_d$$

The TLM algorithm for diffusion/drift

The principle of superposition shows that in order to model drift the addition of the current generator to the TLM one-dimensional model produces additional voltage at the node of

$$I(x)\left[\frac{R+Z}{2}\right] \tag{7.25}$$

In the modified algorithm for drift and diffusion the nodal voltage ${}_k\phi_m$ is defined in terms of incident pulses and the mean difference between neighbouring nodes ϕ_{x+1} and ϕ_{x-1} at the previous iteration interval.

$${}_k\phi(x) = ({}_k^iV_L(x) + {}_k^iV_R(x)) + I(x)\left[\frac{R+Z}{2}\right] \tag{7.26}$$

where

$$I(x) = g_m\left[\frac{{}_{k-1}\phi(x+1) - {}_{k-1}\phi(x-1)}{2}\right]$$

$$\begin{aligned} {}_k^sV_L(x) &= \frac{1}{R+Z}\left[Z_k\phi(x) - (R-Z)\,{}_k^iV_L(x)\right] \\ {}_k^sV_R(x) &= \frac{1}{R+Z}\left[Z_k\phi(x) - (R-Z)\,{}_k^iV_R(x)\right] \end{aligned} \tag{7.27}$$

Scattered pulses travel during time Δt to become incident pulses on adjacent nodes for the next iteration.

Applications of TLM diffusion/drift algorithms

The algorithms describes above have been used very successfully in a range of applications

(1) Modelling of chromatographic and electrophoretic processes [7.7]

A detailed introduction to chromatography can be found many textbooks on analytical chemistry (e.g. Fritz & Schenk [7.8]). Adsorption is a physical process which results in the attraction of atoms or molecules to a surface. It arises as a result of the intermolecular forces between the surface (adsorbant) and the attracted species (adsorbate). The polarity of molecules is an important factor in this process. In any treatment of this phenomenon one can also define a rate of adsorbtion and one can also define a rate of desorbtion (the rate at which species is liberated from a surface). By direct analogy with the law of mass action as it relates to chemical equilibria, one can then define an adsorbtion isotherm as the ratio of the concentration of adsorbed species and the concentration of free species. This is the normal description parameter and chromatography and electrophoresis depend in the first instance on different species having different adsorbtion isotherms.

The second factor is the injection of a mixture of species as a delta-like impulse into a flowing solvent or eluent. In this state the solutes undergo diffusion and adsorbtion which eventually results in species separation. Since the chromatographic process depends on a very large number of variables which are often difficult to reproduce in different systems and/or at different times it is convenient to define a parameter called the retention time (the time a solute to travel a fixed distance) and in any qualitative analysis this is compared against the retention time for one or more standard materials. Standards are also used in quantitative analysis which is based upon the area under the solute curve versus time curve which gives total concentration.

Many molecules such as amino acids display an end-to-end dipole or zwitter-ion effect and the extent of the magnitude of the dipole moment can be used as an analytic signature. Thus when an electrical bias is applied to the system zwitter-ion molecules will migrate in the field. Again the extent of movement as a result of a field can be used in material analysis. A standard arrangement for electrophoresis is to have a two dimensional system whereby the movement of solvent/solute in the z (vertical) direction is due to capillary action and the movement in the x (lateral) direction is due to the electric field.

If paper or fine silica is used as the adsorbant, then the extent of separation makes it possible to physically isolated sections and subsequently extract individual species. To this extent chemists and biochemists utilise these phenomena as preparative tools.

Before we develop a TLM description we must first define a parameter which describes the effects of adsorbtion. The Langmuir adsorbtion isotherm [7.9] as the simplest example. This states that the concentration of immobilised diffusant (S) is directly proportional to the concentration of free diffusant (C).

$$S = KC \tag{7.28}$$

K is a constant called the capacity ratio. When diffusion is accompanied by adsorbtion the diffusion equation becomes [7.10]:

$$\frac{dc}{dt} = D\frac{d^2C}{dt^2} - \frac{dS}{dt} \tag{7.29}$$

Using the derivative of the adsorbtion isotherm the previous equation becomes:

$$\frac{dc}{dt} = D\frac{d^2C}{dt^2} - k\frac{dC}{dt} \tag{7.30}$$

In a rearrangement to express Fick's second law of diffusion this gives:

$$\frac{\partial C}{\partial t} = \left[\frac{D}{1+K}\right]\frac{\partial^2 C}{\partial x^2} \tag{7.31}$$

so that in the presence of adsorbtion the diffusion constant becomes

$$D_{\text{eff}} = \frac{D}{1+K} \tag{7.32}$$

The adsorbtion isotherm and therefore the effective diffusivity is a property of a material and substrate combination. Using this approach it was possible to achieve separation of species and it was also possible to demonstrate that the extent of separation was improved by using a larger flow-rate.

The algorithms for electrophoresis are very similar to those which are covered in the next application.

(2) Simulation of Minority Carrier behaviour in an 'Infinite' Semiconductor

Al-Zeben *et al* [7.6] have shown that a TLM description gives a near perfect match with the equation for minority carrier diffusion, drift and recombination which was demonstrated experimentally by Haynes and Shockley [7.11]. The thermal equilibrium of charge carriers in a semiconductor can be disturbed if excess minority carriers are introduced. Equilibrium is restored during an interval of time by a process called recombination. During this time the excess carriers will diffuse and drift within an electric field, if present. The one-dimensional behaviour is described by:

$$\frac{\partial^2 p}{\partial x^2} - \frac{\mu E_x}{D}\frac{\partial p}{\partial x} - \frac{p}{D\tau} = \frac{1}{D}\frac{\partial p}{\partial t} \tag{7.33}$$

E is the electric field. D and μ are the minority carrier diffusion constants and mobility, τ is the minority carrier life time and p is the concentration of holes in the semiconductor.

The circuit shown in figure 7.3 is similar to the one in figure 7.1 with the addition of a leakage resistor R_L connected across the node. It will be shown that while the current generator models drift (as before) the resistor models the recombination term in the transport equation.
Voltage analysis gives

$$-\frac{\partial V_x}{\partial x} = R_d I_x + L_d \frac{\partial I_x}{\partial t} \tag{7.34}$$

or if Ld $\ll R_d$ this becomes

$$\frac{\partial V_x}{\partial x} = -R_d I_x \tag{7.35}$$

Analysis of the current through the capacitor $C_d \Delta x$ gives

$$\frac{\partial I_x}{\partial x}\Delta x = I_m - \frac{V_x}{R_L} - C_d \Delta x \frac{\partial V_x}{\partial t} \tag{7.36}$$

By taking $G_L = 2/R_L$ and $g_m \Delta V_x$ where ΔV_x is the mean voltage difference between two adjacent nodes or $\Delta V_x = (V(x+1) - V(x-1))/2$ gives

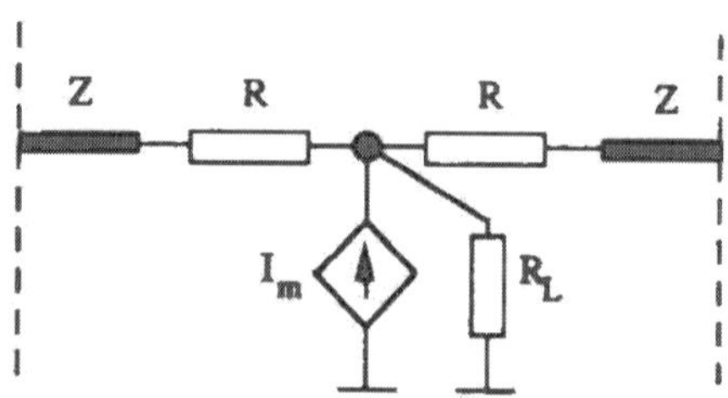

Figure 7.3 A diffusion/drift node with a shunt loss resistor.

$$\frac{\partial I_x}{\partial x}\Delta x = g_m \Delta V_x - G_L V_x - C_d \Delta x \frac{\partial V_x}{\partial t} \tag{7.37}$$

or

$$\frac{\partial I_x}{\partial x} = g_m \frac{\partial V_x}{\partial x} - G_d V_x - C_d \frac{\partial V_x}{\partial t} \tag{7.38}$$

where $G_d = G_L/\Delta x$ or the distributed leakage conductance across the node.

Combining equations gives

$$\frac{\partial^2 V}{\partial x^2} + g_m R_d \frac{\partial V}{\partial x} + G_d R_d V_x = R_d C_d \frac{\partial V}{\partial t} \tag{7.39}$$

Comparing this with the transport equation (eqn. (7.33)) yields the following equivalence:

$$\begin{aligned} D &\equiv \frac{1}{R_d C_d} & p &\equiv V_x \\ \tau &\equiv \frac{C_d}{G_d} & \frac{\partial p}{\partial x} &\equiv -I_x R_d \\ -\mu E_x &\equiv \frac{g_m}{C_d} \end{aligned} \tag{7.40}$$

The semiconductor parameters can now be expressed in terms of model parameters as:

$$\begin{aligned} \frac{R}{Z_0} &= \frac{\Delta x^2}{2D\Delta t} \\ \frac{R_1}{Z_0} &= \frac{\tau}{\Delta t} \\ \frac{g_m}{Y_0} &= \frac{-\mu E}{\Delta x/\Delta t} \end{aligned} \tag{7.41}$$

The algorithm is in almost all cases exactly as previously except that allowances must be made for the leakage resistor, R_L in the nodal potential.

$$\begin{aligned} {}_k\phi(x) &= \frac{2R_L}{2R_L + R + Z_0}\left[\left({}_k^iV_L(x) + {}_k^iV_R(x)\right) + {}_kI(x)\frac{R+Z_0}{2}\right] \\ &\text{where} \quad {}_kI(x) = g_m\left[\frac{{}_{k-1}\phi(x+1) - {}_{k-1}\phi(x-1)}{2}\right] \end{aligned} \tag{7.42}$$

The model has been tested and gives an almost perfect fit with the analytical solution for the range of conditions that were tested.

(3) Simulation of minority carrier behaviour in a finite semiconductor

Saleh and colleagues [7.12] then extended this further to consider the case where there is a finite boundary. However, the presence of a metallic collector contact influences the carrier distribution, which can no longer be described by equation (7.33). The physical problem now becomes a finite length semiconductor, subject to the boundary conditions describing surface recombination at the contact. Duggan and Berz [7.13] characterised the effect of the contact by introducing a finite surface recombination velocity, S, which describes the hole current $J(x,t)$ at the contact as:

$$J(0,t) = -\mu E_x p + D\left.\frac{\partial p}{\partial x}\right|_{x=0} = Sp(0,t)$$

or

$$\left.(S+\mu E_x)p\right|_{x=0} = D\left.\frac{\partial p}{\partial x}\right|_{x=0} \tag{7.43}$$

The analytical solution to such problems cannot be obtained and at present computational methods provide the only feasible approach.

The scattering and connect processes are as before, but on this occasion we require a re-definition of the boundary. Using the equivalence presented above the boundary condition takes the new form:

$$\left.\left(S-\frac{g_m}{C_d}\right)V\right|_{x=0} = -\left.\frac{I}{C_d}\right|_{x=0} \tag{7.44}$$

Since the voltage across the boundary resistor $V_B = V|_{x=0}$ and the current through the resistor $I_B = -I|_{x=0}$, therefore

$$\left(S-\frac{g_m}{C_d}\right)V_B = \frac{I_B}{C_d}$$

or

$$R_B = \frac{V_B}{I_B} = Z_0\,\frac{\Delta x/\Delta t}{(S+\mu E_x)} \tag{7.45}$$

The scattered voltage pulse to the left of the first node at the k th interval ${}_k^sV_L$ encounters the resistor R_B and partially reflects to become incident on the same node from the left at the next iteration or

$$ {}_{k+1}^{\;i}V_L(1) = \frac{R_B - Z_0}{R_B + Z_0}\, {}_k^s V_L(1) \tag{7.46} $$

This equation modifies the connect process for the first node. The scattering algorithm for the first node is also modified by replacing $g_m\,({}_{k-1}\phi(x+1) - {}_{k-1}\phi(x-1))/2$ with $g_m\,({}_{k-1}\phi(1) - {}_{k-1}\phi(2))/2$.

(4) TLM modelling of heat exchangers

In this treatment we consider a heat exchanger as a system of counter-flowing fluids (one initially hot and the other initially cold) which are separated by a thin metallic membrane. This simple structure presents the modeller with some interesting challenges. In the initial stage and during any thermal transients it is a moving boundary, drift diffusion problems that is numerically stiff; that is the time constants of the metal interface and fluids are likely to be sharply different. A numerical treatment using the time-step appropriate to the metal will be computationally wasteful, while the use of the fluid time-step might introduce inaccuracies.

The application of TLM to this system was first studied by Lee [7.14] using the crude discrete transfer method. In spite of the problems associated with this approach he did manage to demonstrate that maximum heat transfer was maintained when the flow rates were sufficiently high as to minimise heat transfer into the fluids perpendicular to the metallic interface.

de Cogan and Chichlowski [7.15] used a biased scattering coefficient approach. For a two-dimensional formulation the scattering equation is given by:

$$ \begin{bmatrix} {}^sV_L \\ {}^sV_R \end{bmatrix} = \begin{bmatrix} \rho-\alpha & \tau-\alpha \\ \rho+\alpha & \tau+\alpha \end{bmatrix} \begin{bmatrix} {}^sV_L \\ {}^sV_R \end{bmatrix} \tag{7.47} $$

Acting on a suggestion of Johns [7.16] they started the simulation process at a time-step appropriate to the metal membrane. They confirmed that there was little loss in accuracy if as the temperature distribution in the metal approached equilibrium, the time-step was progressively changed to a value which was representative of the fluid thermal time-constant.

The stability and accuracy of diffusion-drift algorithms

More recently Chakrabarti [7.17] has identified some problems with TLM algorithms for drift-diffusion. He was attempting to use these to simulate carrier transport in semiconductor devices where there can be very significant field gradients. Contrary to the normal situation with TLM there were conditions when the algorithms appeared to become unstable. At first this was thought to be due to the fact that the current generator is calculated as in eqn (7.42), i.e. using values of neighbouring nodes calculated at the previous time-step. This idea was first tested by comparing the situation for a semi-implicit algorithm. ${}_k\phi(x)_1$ was initially calculated using ${}_kI(x)$ as defined above. These values were then used to obtain an improved estimate for the current,

$$ {}_kI(x)_2 = g_m \left[\frac{{}_k\phi(x+1)_1 - {}_k\phi(x-1)_1}{2}\right] \tag{7.48}$$

This was used for an improved estimate of the potentials ${}_{k+1}\phi(x)_2$ and the process was repeated until convergence was obtained in the calculation

$$ {}_k\phi(x)_\infty = {}^i_kV_L(x) + {}^i_kV_R(x) + g_m\left[\frac{R+Z}{2}\right]\left[\frac{{}_k\phi(x+1)_\infty - {}_k\phi(x-1)_\infty}{2}\right] \tag{7.49}$$

The end result of this approach appeared to be successful, but the results within the universally stable region did not always appear to be as accurate as the Saleh method. For that reason a fundamental reappraisal was undertaken which included an assessment of the consistency, stability and accuracy of TLM algorithms for drift-diffusion and its relationship to the equivalent explicit finite difference algorithm.

In the case of the finite difference formulation, we start with

$$ \begin{aligned} {}_{k+1}\phi(x) &= {}_k\phi(x) - \left[\frac{\mu E\Delta t}{2\Delta x}\right]\{{}_k\phi(x+1) - {}_k\phi(x-1)\} \\ &\quad + \left[\frac{D\Delta t}{\Delta x^2}\right]\{{}_k\phi(x+1) + {}_k\phi(x-1) - 2{}_k\phi(x)\} \end{aligned} \tag{7.50}$$

Each side is expanded as a Taylor series. Using cancellation followed by division by Δt we obtain:

$$ \begin{aligned} \phi_t &= -\mu E\phi_x + D\phi_{xx} - \frac{1}{2}\,\phi_{tt}\Delta t - \frac{1}{6}\,\phi_{ttt}\Delta t^2 \ldots \\ &\quad - \frac{1}{6}\,\mu E\phi_{xxx}\Delta x^2 \ldots + \frac{1}{12}\,D\phi_{xxxx}\Delta x^2 \ldots \end{aligned} \tag{7.51}$$

The subscripts t and x refer to derivatives with respect to space or time.

This can be handled more easily if the above equation undergoes a process of repeated differentiation and back substitution with respect to time, so that the right side only contains spatial derivatives.

$$\phi_t = -\mu E\phi_x + D\phi_{xx} - \left\{ (\mu E)^2 \Delta t \frac{\phi_{xx}}{2} + \left[-\frac{\mu E \Delta x^2}{6} + \mu E D \Delta t - \frac{(\mu E)^3 \Delta x^3}{3} \right] \phi_{xxx} + \cdots \right\} \tag{7.52}$$

The terms inside the curly brackets are truncation error terms, but they do tend to zero as $\Delta x, \Delta t \to 0$.

The same approach can be used for the conventional explicit TLM. The expression for the nodal potential (eqn. 7.26) can be expressed as:

$${}_k\phi(x) = {}^i_kV_L(x) + {}^i_kV_R(x) + B[{}_{k-1}\phi(x+1) - {}_{k-1}\phi(x-1)] \tag{7.53}$$

where $B = (g_m/4)[R + Z_0]$.

Now as it is undesirable to have incident voltage terms we can back substitute in time

$$\begin{aligned} {}^i_kV_L(x) &= {}^r_{k-1}V_R(x-1) = (\rho - \tau)\, {}^i_{k-1}V_R(x-1) + \tau\, {}_{k-1}\phi(x-1) \\ {}^i_kV_R(x) &= {}^r_{k-1}V_L(x+1) = (\rho - \tau)\, {}^i_{k-1}V_L(x+1) + \tau\, {}_{k-1}\phi(x+1) \end{aligned}$$

going further backwards in time

$$\begin{aligned} (\rho - \tau)\, {}^i_{k-1}V_L(x+1) &= (\rho - \tau)\, {}^r_{k-2}V_R(x) \\ &= (\rho - \tau)^2\, {}^i_{k-2}V_R(x) + (\rho - \tau)\tau\, {}_{k-2}\phi(x) \end{aligned}$$

so

$${}^i_kV_L(x) = (\rho - \tau)^2\, {}^i_{k-2}V_R(x) + (\rho - \tau)\tau\, {}_{k-2}\phi(x) + \tau\, {}_{k-1}\phi(x+1) \tag{7.54}$$

similarly,

$$\begin{aligned} (\rho - \tau)\, {}^i_{k-1}V_R(x-1) &= (\rho - \tau)\, {}^r_{k-2}V_L(x) \\ &= (\rho - \tau)^2\, {}^i_{k-2}V_L(x) + (\rho - \tau)\tau\, {}_{k-2}\phi(x) \end{aligned}$$

so

$${}^i_kV_R(x) = (\rho - \tau)^2\, {}^i_{k-2}V_L(x) + (\rho - \tau)\tau_{k-2}\phi(x) + \tau_{k-1}\phi(x-1) \tag{7.55}$$

When all of these substitutions have been made we obtain an expression which has much in common with eqn (7.13a).

$$
\begin{aligned}
{}_{k+1}\phi(x) = (\rho^2 - \tau^2)_{k-1}\phi(x) + B(\rho - \tau)^2[{}_{k-2}\phi(x+1) - {}_{k-2}\,\phi(x-1)] \\
+ (\tau + B)_k\phi(x+1) + (\tau - B)_k\phi(x-1)
\end{aligned}
$$

Now, we can expand each of these as a Taylor series:

When this has been simplified we obtain ϕ_t

$$
\phi_t = \phi_x\left\{\left[\frac{2B\Delta x}{\Delta t}\right]2\tau\right\} + \phi_{xx}\left\{\left[\frac{x^2}{\Delta t}\right]\frac{\tau}{2\rho}\right\} + \phi_{xt}\left\{4\Delta x B\left[\frac{\rho-\tau}{2\rho}\right]\right\} - \phi_{tt}\left\{\frac{\Delta t\rho}{2\tau}\right\} \tag{7.56}
$$

The TLM coefficients are then substituted by their physical equivalents and the equation reduces to:

$$
\phi_t = -\mu E\phi_x + D\phi_{xx} + \mu E\left\{\Delta t - D\frac{\Delta t^2}{\Delta x^2} - \frac{\Delta x^2}{4D}\right\}\phi_{xt} - D\frac{\Delta t^2}{\Delta x^2}\phi_{tt} \tag{7.57}
$$

Thus the truncation errors do present a consistency problem unless care is taken in the manner that Δx and Δt tend to zero.

Using the same process as in the finite difference treatment above we can have spatial derivatives only on the right hand side.

$$
\begin{aligned}
\phi_t = -\mu E\phi_x + D\phi_{xx} + \Bigg\{(\mu E)^2\left[\frac{\Delta x^2}{4D} - \Delta t\right]\phi_{xx} \\
+\mu E\left[D\Delta t - \frac{\Delta x^2}{4}\right]\phi_{xxx} + \cdots\Bigg\}
\end{aligned} \tag{7.58}
$$

The truncation terms which contain even orders of derivatives (i.e. ϕ_{xx}, ϕ_{xxxx}) lead to what is called *implicit numerical diffusion* and the odd order derivatives lead to *implicit numerical dispersion*.

A similar treatment for a semi-implicit TLM algorithm yields

$$
\phi_t = -\mu E\phi_x + D\phi_{xx} + \left\{(\mu E)^2\left[\frac{\Delta x^2}{4D}\right]\phi_{xx} - \mu E\left[\frac{\Delta x^2}{4}\right]\phi_{xxx} + \ldots\right\} \tag{7.59}
$$

In this case it will be noted that any influence from Δt is absent. All of these errors are now summarised in Table 7.2.

Table 7.2

	Diffusion error	*Dispersion error*
Explicit finite difference	$-\frac{(\mu E)^2}{2}\Delta t$	$-\frac{\mu E\Delta x^2}{6} + \mu E\Delta t - \frac{(\mu E\Delta t)^3}{3}$
Explicit TLM	$(\mu E)^2\left[\frac{\Delta x^2}{4D} - \Delta t\right]$	$\mu E\left[D\Delta t - \frac{\Delta x^2}{4}\right]$
Semi-implicit TLM	$(\mu E)^2\left[\frac{\Delta x^2}{4D}\right]$	$-\mu E\left[\frac{\Delta x^2}{4}\right]$

The errors due to numerical diffusion and dispersion can be observed in different ways. We can look at how choice of diffusion and convection numbers determine the implicit numerical errors. The different algorithms can also be investigated using the standard von Neumann stability analysis. This represents the initial condition as a series of Fourier modes and checks whether the solution remains bounded. An expression was developed for the amplification factor, ξ and the conditions when $|\xi| \leq 1$ were then investigated. This can be summarised within the Courant-Friedrichs-Lewy stability criterion [7.19] which states that diffusion-convection algorithms are stable provided the following condition is satisfied:

$$1 \geq 2r \geq s^2 \tag{7.60}$$

(where s is the convection number and r is the diffusion number)

This information can be summarised as the spectral dependence of amplification factor and this is shown for the explicit finite difference and TLM in figure 7.4.

Now that the question of stability of TLM drift-diffusion algorithms has been put on a sound basis we can return to the question of the errors. Table 7.2 summarised the diffusion and dispersion errors and these are now presented graphical form. Figure 7.5 compares the diffusive errors plotted against convection number for a range of diffusion numbers. In a similar way, the numerical dispersion errors as a function of r and s are shown in figure 7.6. It can be seen that there is a value of diffusion number, $D\Delta t/\Delta x^2 = 0.25$ when diffusive and dispersive errors are negligible for explicit TLM.

The effects of implicit numerical diffusion and dispersion can now be demonstrated by considering a one-dimensional drift-diffusion model of charge carriers. In this simulation charge was injected and allowed to drift/diffuse for 0.5μs. The convection number was maintained constant (0.5). The space and time discretisations were varied so as to have different diffusion numbers. In each of the two cases which were considered the results for explicit finite difference and TLM were compared with the analytical result. In the first case (figure 7.7(a)) where $r = 0.35$ there is *infusion*; the level of diffusion is less than the analytical value. When $r = 0.15$ (figure 7.7(b)) the situation is quite different. TLM displays extra diffusion while finite difference is still infusive when compared with the analytical result.

In concluding this chapter we can also present the stability data of figure 7.4 in an alternative format which is shown in figure 7.8. This is particularly useful in determining whether an algorithm will be stable for a given set of discretisation and field conditions.

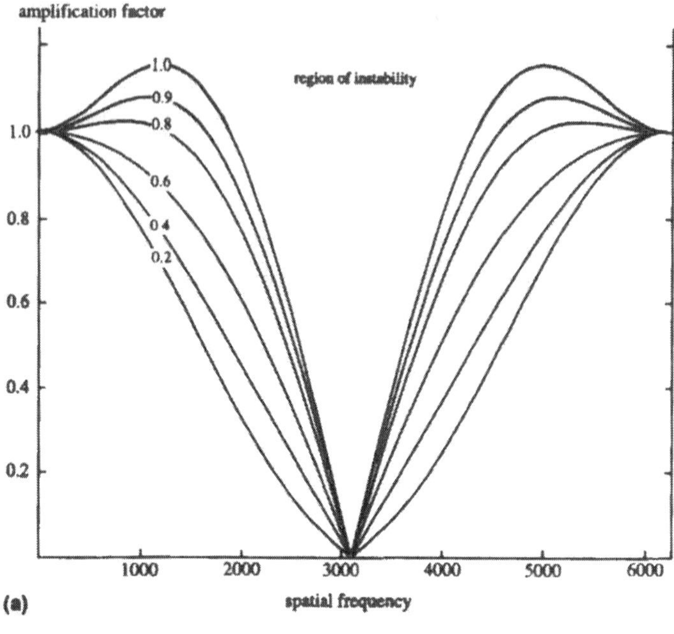

(a)

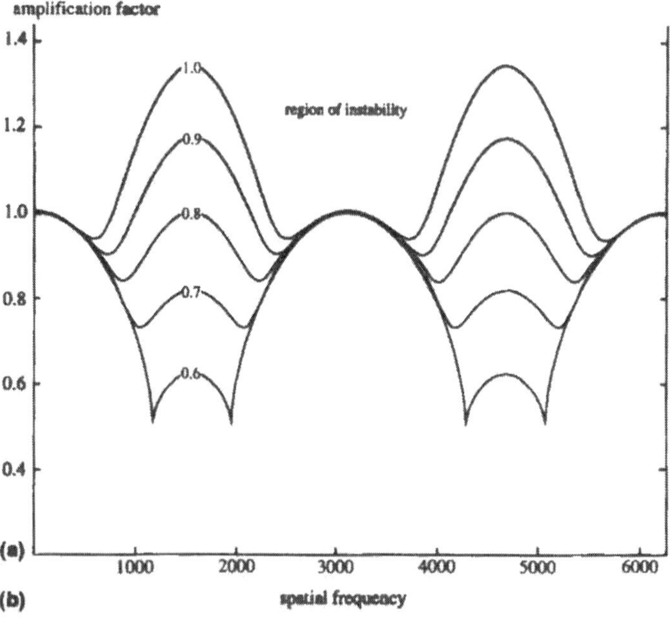

(b)

Figure 7.4 Amplification factor versus spatial frequency as a function of convection number: (a) explicit finite difference, (b) explicit TLM.

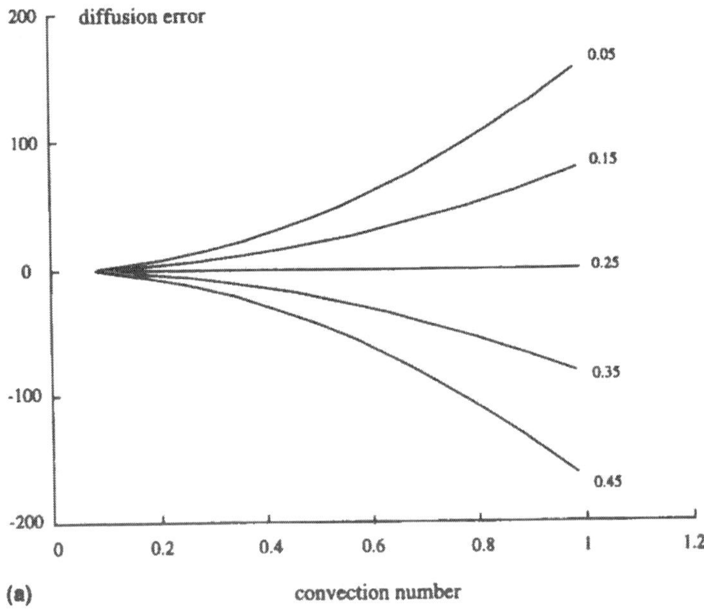

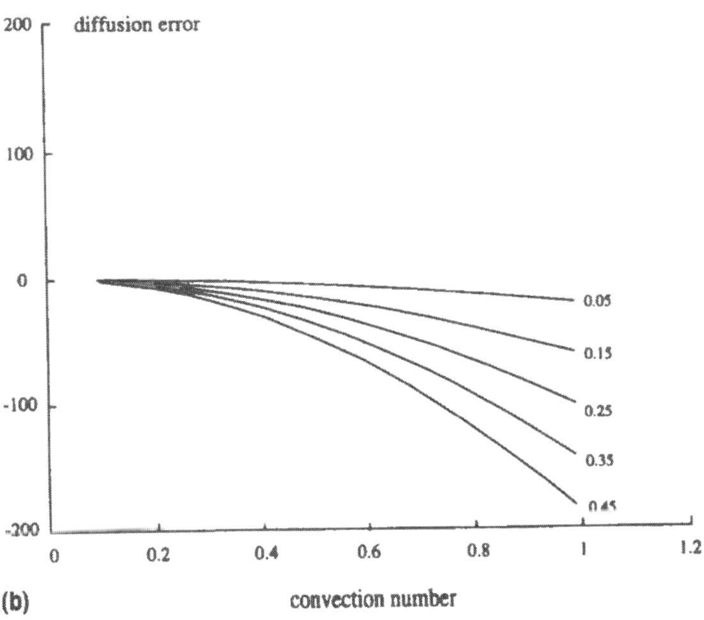

Figure 7.5 Diffusive errors against convection number as a function of diffusion number: (a) TLM (explicit and implicit), (b) explicit finite difference.

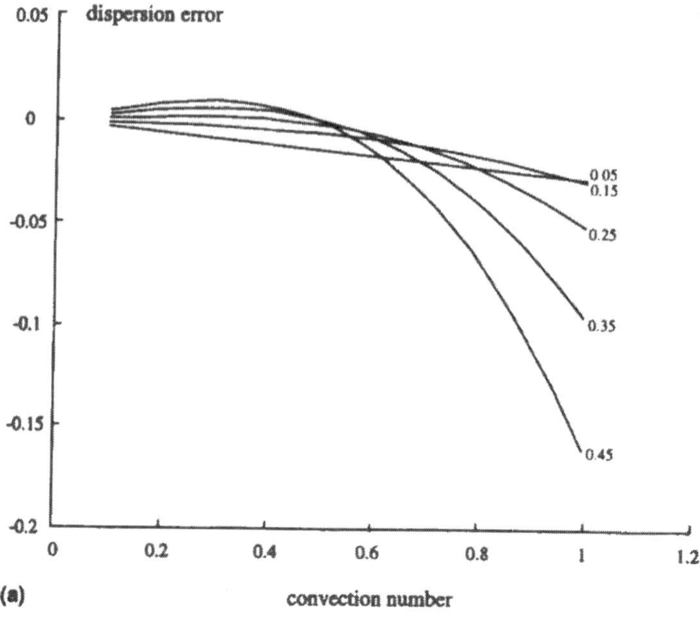

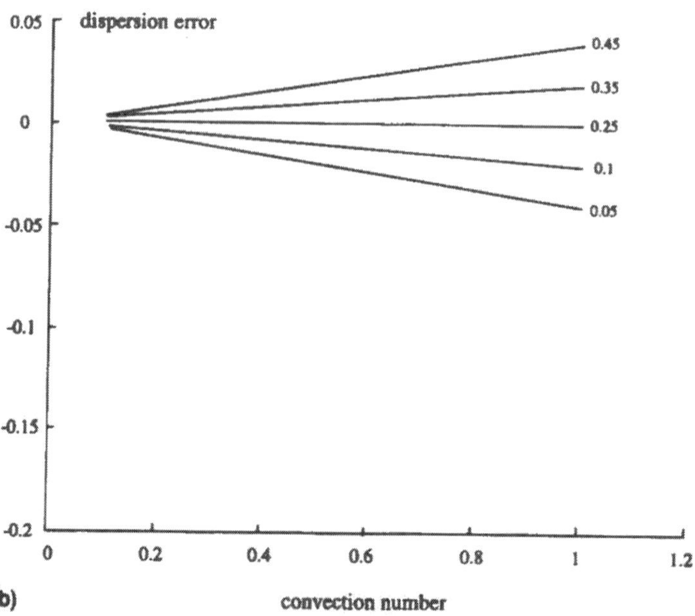

Figure 7.6 Dispersive errors against convection number as a function of diffusion number: (a) TLM (explicit and implicit), (b) explicit finite difference.

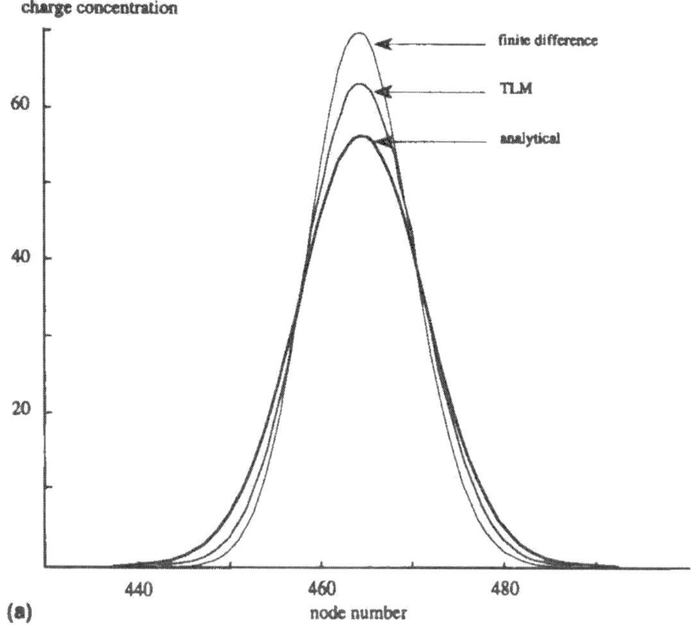

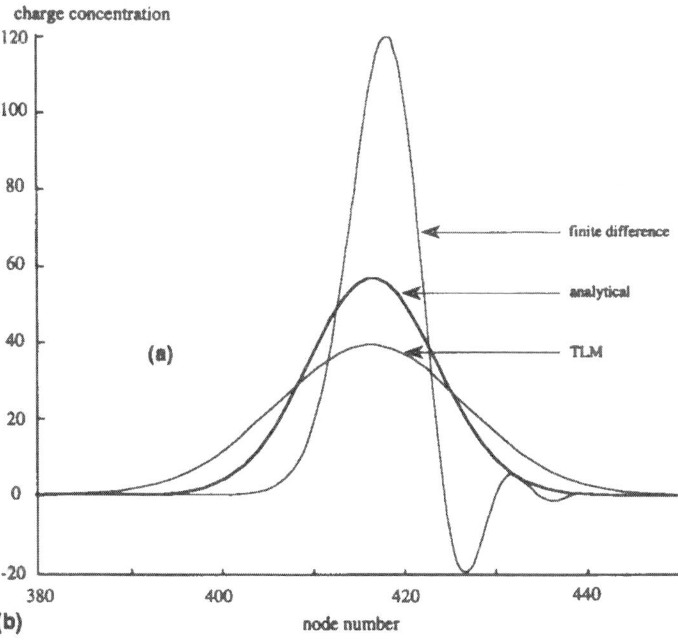

Figure 7.7 A comparison of the concentration profiles in a one-dimensional drift-diffusion after 0.5 μs with constant convection number. (a) diffusion number = 0.35, (b) diffusion number = 0.15.

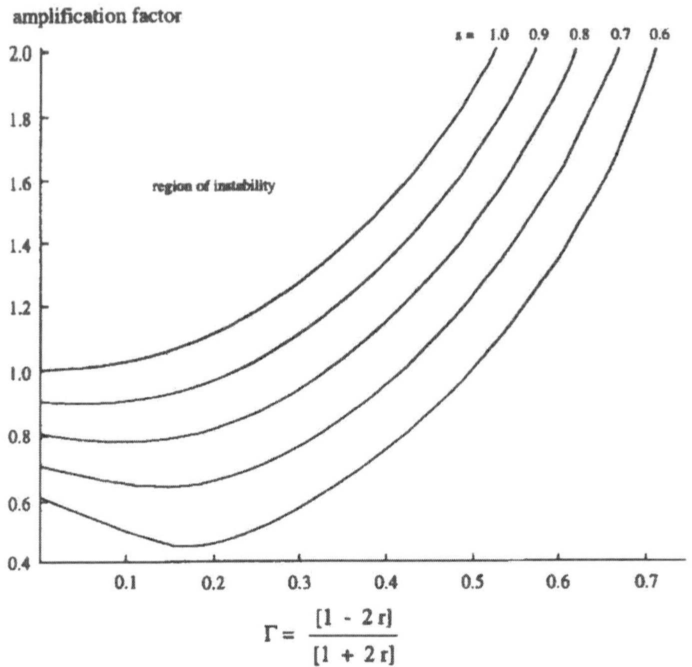

Figure 7.8 ξ versus TLM diffusion number as a function of convection number.

References

[7.1] P. Enders, Queuing and discrete modelling of drift-diffusion problems, (unpublished manuscript).

[7.2] P. Enders, TLM techniques for the modelling of diffusion-drift phenomena, (unpublished manuscript).

[7.3] T. Sato and K. Terao, An interpretation of the telegraph equation for animal movement near a boundary wall, *Japanese Journal of Applied Physics* **25**, (1986) L229 – L302.

[7.4] A. Smith, Transmission Line Matrix modelling, optimisation and application to adsorption phenomena, *B. Eng. Thesis*, Nottingham University, 1988.

[7.5] D. de Cogan and M. Henini, Transmission line matrix (TLM): a novel technique for modelling reaction kinetics, *J. Chem. Soc., Faraday Trans.*, 2, **83**, 8843 – 8855, (1987).

[7.6] M.Y. Al-Zeben, A.H.M. Saleh and M.A. Al-Omar, TLM modelling of diffusion, drift and recombination of charge carriers in semiconductors, *International Journal of Numerical Modelling* **5**, (1992), 219 – 225.

[7.7] D. de Cogan and A.H.M. Saleh, TLM modelling of chromatography and electrophoresis, (unpublished manuscript in the possession of this author).

[7.8] J.S. Fritz and G.H. Schenk, "Quantitative Analytical Chemistry", Allyn and Bawn, USA, 1987, chapter 21.

[7.9] S. Glasstone, "A Textbook of Physical Chemistry", Macmillan London 1951, 1198.

[7.10] J. Crank, N.R. McFarlane, J.C. Newby, G.D. Paterson and J.B Pedley "Diffusion Processes in Environmental Systems", Macmillan London 1981, 104.

[7.11] J.R. Haynes and W. Schockley, The mobility and life of injected holes and electrons in germanium, *Phys.Rev.*, 81, (1951) 835.
[7.12] A.H.M Saleh, M.Y. Al-Zeben, TLM modelling of the Haynes-Shockley experiment in the presence of a metallic contact, (a copy of the manuscript which had not been submitted for publication before the death of A.H.M. Saleh is in the possession of this author).
[7.13] G. Duggan and F. Berz, "The theory of the Shockley-Haynes experiment: contact effects", *J. Appl. Phys.* 53, (1982) 470–476.
[7.14] K.S.T. Lee, TLM modelling of heat exchangers, *B. Eng. thesis*, Nottingham University, 1987.
[7.15] D. de Cogan and K.O. Chichlowski, TLM modelling of thermal transfer in heat exchangers, *Proc. Eurotherm 36* (Advanced Concepts and Techniques in Thermal Modelling) Elsevier 1996 pp. 247–252 (ISBN 2-906077-77-1).
[7.16] P.B. Johns (private discussions with the author in 1987).
[7.17] A. Chakrabarti, Transmission Line Matrix modelling for semiconductor transport, *PhD thesis*, UEA Norwich 1996.
[7.18] R. Courant, K. Friedrichs and H. Lewy, On the partial difference equations of mathematical physics, *IBM Journal* (1967) 215–234.

Chapter 8

TLM algorithms for Laplace and Poisson equations

Introduction

The Laplace equation is without doubt one of the most important equations of physics. It arises naturally in many branches of mathematics where it is a member of the class of elliptic equations. It is one of the cornerstones of potential theory and in electromagnetics it describes the potential distribution in a region of charge-free space.

$$\nabla^2 V = 0 \tag{8.1}$$

In terms of diffusion theory it can be interpreted as an equilibrium state: i.e. Fick's second law of diffusion with the time derivative set to zero:

$$D\nabla^2 C(x, y, z, t) = \frac{\partial}{\partial t} C(x, y, z, t) = 0 \tag{8.2}$$

In heat-flow it implies that there is a fixed temperature gradient and therefore a constant heat-flux.

In the presence of a distribution of fixed heat sources the thermal diffusion equation can be written as:

$$D\nabla^2 T(x, y, z, t) = \frac{\partial}{\partial t} T(x, y, z, t) - \Theta(x, y, z) \tag{8.3}$$

where $\Theta(x, y, z)$ is the temperature arising from heat generated per unit time due to the sources within a unit volume of space. As the system approaches equilibrium we obtain another important elliptic equation: Poisson's equation.

$$\nabla^2 T(x, y, z, t) = -\frac{\Theta(x, y, z)}{D} \tag{8.4}$$

In potential theory this is used to express the effects of a time invariant spatial distribution of charge

$$\nabla^2 V = -\frac{\rho}{\varepsilon_r \varepsilon_0} \tag{8.5}$$

Poisson's equation is particularly important in semiconductor transport modelling. In fully depleted semiconductors the potential due to a distribution of fixed donor ions $N_d(x, y, z)$ gives rise to $\nabla^2 V = -eN_d/\varepsilon_r\varepsilon_0$ which is subject to boundary conditions such as the potentials applied to the external contacts. The solution yields an electric field which is proportional to distance. It can be applied to more complicated situations: for example a semiconductor junction in the presence of high carrier injection, where the local concentration of charge is a function of the local field.

$$\frac{d^2 V(x)}{dx^2} = -\frac{e}{\varepsilon_r \varepsilon_0} N_d(x) - \frac{J(x)}{\mu \varepsilon_r \varepsilon_0} \left[\frac{dV(x)}{dx}\right]^{-1} \tag{8.6}$$

So long as the dielectric relaxation time is very small it can also be used in applications where the charge distribution changes with time. In this context 'very small' means that the time for the potential distribution to reach equilibrium is very short by comparison with the time variations of the charge distribution.

Many of these problems are only amenable to numerical solution and finite difference, finite element and Monte Carlo methods are widely used. This chapter outlines the first application of TLM techniques to Laplace and Poisson problems.

TLM algorithms for the Laplace equation

de Cogan *et al* [8.1] investigated the conditions when a TLM diffusion algorithm was allowed to approach equilibrium. Because of this there was no requirement to work with specific values of reflection and transmission coefficients. The objective was to obtain a solution of the required accuracy in the fastest possible time. This latter requirement might be specially important if a Laplace or Poisson solver is just one part of a bigger system such as semiconductor charge transport modelling with time-varying conditions.

The test-piece was a one-dimensional sample with a potential of 100V applied at one side and the other side held at ground. In the first instance iterations on a link-line network were continued until the potential at the mid-point in the problem space was within 10^{-3}% of the true value. The relationship between the value of ρ and the number of iterations to obtain this accuracy was investigated. This was later changed to a requirement to have all nodes in the problem space at 10^{-3}% accurate or better.

It was found that there was a value of ρ (termed ρ_{opt}) around which the number of iterations necessary to achieve the required accuracy increased significantly. It was noted that if ρ was less than ρ_{opt} then there was oscillatory convergence. If ρ was greater than ρ_{opt} then the convergence was significantly slower.

Several other observations arose. It was noted that the value of ρ_{opt} decreased as the number of node points, n was increased. In the vicinity of the optimum value the number of iterations dropped suddenly so that at ρ_{opt} with an accuracy of 10^{-3}% the value of $k_{min}\,\Delta x$ (k_{min} is the minimum number of iterations) was approximately four times the number of nodes in the problems space. This can be interpreted as meaning that at ρ_{opt} the input data from the terminals require four transits of the mesh in order to establish equilibrium.

Some of these results can be seen in Table 8.1 along with the number of iterations to convergence and the value of τ_{opt}/ρ_{opt}, which is in fact a direct measure of the optimum $\Delta t/(R_d\,C_d\,\Delta x^2)$. It can be seen that this value is approximately one tenth of the number of iterations required for convergence which suggests that it could be used for determining optimum conditions. Table 8.1 also includes the number of iterations which were required to obtain the same accuracy in a simple random walk (SRW) model of the same system. It is worth noting that there was a significant improvement when an optimised successive over-relaxation (SOR) finite difference formulation was used but the convergence was still inferior to that of the optimised TLM algorithm.

Table 8.1

n	ρ_{opt}	k_{min}	(τ_{opt}/ρ_{opt})	k(SRW)
5	0.349	20	1.865	59
7	0.298	26	2.355	119
9	0.250	35	3.0	197

Although the four transits concept can be a useful metric, there is a need for the modeller to experiment as the number of iterations to convergence varies quite steeply about ρ_{opt}. Taking as an example the situation at $n=9$, where $\rho_{opt}=0.25$ and $k_{min}=35$. It was found that if $\rho=0.23$ then $k=42$ while if $\rho=0.27$ then $k=55$.

Experiments were extended to higher levels of accuracy. It was found that if the accuracy level was increased by a factor 10 to 10^{-4}% then five mesh transits were required. Similarly six transits were required for 10^{-5}% accuracy.

Some currently unpublished work on two-dimensional Laplace problems suggests that there are very sharply defined error surfaces and that the fastest convergence is obtainable for a given accuracy when the reflection coefficient approaches that of the lossless wave (i.e. $\rho \rightarrow 0.5$). The exact values appear to depend on the level of discretisation of the problem.

TLM algorithms for the Poisson equation

In the TLM solution of Poisson's equation there is also a choice: whether a link-line or a link-resistor network is used. The effects of the charge can either be introduced at the centre of the node or at the interface between nodes. The treatment presented here is for storage at the centre of a node.

A charge, Q stored by the capacitance of a transmission line will give rise to a voltage (V_Q). This is analogous to heat injection and the treatment is similar. If the charge is moved during the iteration time step Δt then this represents a current (I). In the circuit analysis of a transmission line with a node centred current generator we arrive at the equation:

$$\nabla^2 V = L_d C_d \frac{\partial^2 V}{\partial t^2} + R_d C_d \frac{\partial V}{\partial t} - \frac{2 R_d I}{\Delta x} \tag{8.7}$$

Using the usual arguments to approximate this to the diffusion equation, we can now show that

$$\frac{2 R_d I}{\Delta x} \equiv \frac{\rho}{\varepsilon_r \varepsilon_0} \tag{8.8}$$

Starting with the Thévenin circuit for a one-dimensional link-line node with a current generator at the centre we can deduce the nodal voltage as:

$$_{k+1}\phi(x) = {}_{k+}{}^{i}V_L(x) + {}_{k+}{}^{i}V_R(x) + I(x)\left[\frac{R+Z}{2}\right] \tag{8.9}$$

The usual link-line *scatter* and *connect* processes then apply.

The source of current can also be interpreted as a resistor and this is now derived in the context of a link-resistor node. The behaviour of the node in figure 8.1(a) can be represented by a voltage dependent resistor (R_Q), whose current is given by:

$$I(x) = \frac{\phi(x)}{R_Q(x)} = \frac{V_Q}{Z} = \frac{V_Q}{\Delta t/C} = \frac{Q}{\Delta t} \tag{8.10}$$

An analysis using the equivalent circuit (figure 8.1(b)) gives a value for the potential, $\phi(x)$ at the node.

$$\phi(x) = \frac{\dfrac{2\,{}^iV_L(x)}{Z} + \dfrac{2\,{}^iV_R(x)}{Z}}{\dfrac{1}{Z} + \dfrac{1}{Z} + \dfrac{1}{R_Q(x)}} = \frac{2R_Q(x)[{}^iV_L(x) + {}^iV_R(x)]}{2R_Q(x) + Z} \tag{8.11}$$

At this stage the value of the voltage dependent resistor is not known but it can be derived from eqn. (8.10). A convergent value may need some iteration within each time step, but this has been found to be minimal for static cases.

The presence of $R_Q(x)$ causes scattering at the mid-point of the node and the reflected components are given by:

$${}^rV_L(x) = \phi(x) - {}^iV_L(x) \quad \text{and} \quad {}^rV_R(x) = \phi(x) - {}^iV_R(x) \tag{8.13}$$

These pulses continue for a time $\Delta t/2$ until they reach the resistors at the edge of the node where they undergo conventional scattering which is defined by $\rho = R/(R+Z)$, $\tau = Z/(R+Z)$. The summation of scattered and transmitted (from adjacent nodes) defines the incident pulses for the next iteration.

This TLM routine was demonstrated [8.1] by comparing the results with other techniques. The test-piece comprised a one-dimensional mesh of 5 nodes representing the space between two electrodes. One electrode was at

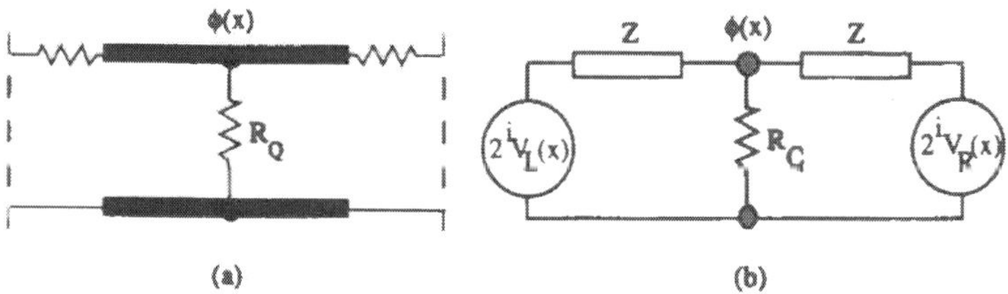

Figure 8.1 (a) Link resistor Poisson node (b) Thévenin equivalent circuit.

100V and the other was at ground potential. A charge of -1 was placed at each point (node edge and node centre). In a finite difference simulation the values were found to converge to the potentials 1, 9, 25, 49 and 81 within 164 iterations. These potentials correspond to the distances L/10, 3L/10, 5L/10, 7L/10 and 9L/10 from the ground electrode, where L was the electrode separation. When successive over-relaxation (SOR) was used the convergence time (to within 10^{-3}% was reduced to 50 iterations. The number of nodes could be increased to any desired size (N_{node}) while altering the nodal charge to $-1(5/N_{\text{node}})^2$.

The TLM simulation was not so straight-forward. The use of half distances ($\Delta x/2$) and half time steps ($\Delta t/2$) in the TLM treatment means that the equivalent charge per node is -4, but even here there is a need for caution. Initial tests gave the correct quadratic behaviour except that the values were each less than the expected by 1 (i.e. 0, 8, 24, 80V). It was realised that there was a difference between the effective charge in the bulk of the test-piece and at a boundary. This can be seen in figure 8.2. The distance between nodes in the bulk is Δx and each bulk node (such as A in figure 8.2) has contributions from a volume 2 Δx (assuming unit area). However, the node nearest the boundary (B in the figure) is set at a distance $\Delta x/2$ from the boundary. The effective charge is then three quarters of the value in the bulk. Once this correction was made then the results converged to the analytical values. These observations can be confirmed using a link-line representation where the potential at the interface between nodes is monitored as the summation of pulses which are incident at the centre of the nodes a half time-step later:

$$ {}_{(k+1)/2}\phi(\text{interface } x, x+1) = {}_{k+\frac{1}{2}}^{\ i}V_R(x) + {}_{k+\frac{1}{2}}^{\ i}V_L(x+1) \qquad (8.14) $$

In this case the nature of the boundary definitions imply 'ghost' half nodes beyond the fixed value boundaries. The volumes at all nodes are therefore identical and this technique also converged to the exact results.

Chakrabarti [8. 2] has pointed out that the inclusion of current generators within a TLM mesh can mean that it is no longer a passive circuit.

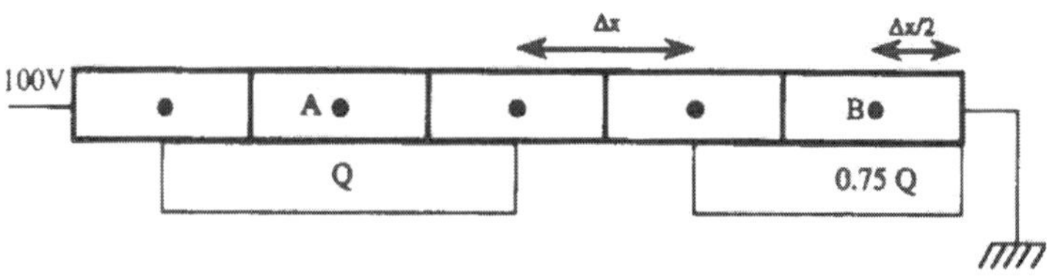

Figure 8.2. Relative volumes of charge (shaded) available for bulk node, A and boundary node, B.

Instability problems might then arise and he demonstrated this as being particularly important in the case of modelling diffusion/drift phenomena (see chapter 7). This aspect was not addressed explicitly, but he did develop expressions for the difference equations of the link-line algorithm.

If we take the nodal voltage at the present moment:

$$_k\phi(x) = {}_k^iV_L(x) + {}_k^iV_R(x) + I(x)\left[\frac{R+Z}{2}\right] \tag{8.14}$$

we can substitute for the incident voltages to get:

$$\begin{aligned} {}_k^iV_R(x) + {}_k^iV_L(y) &= \Gamma[({}_{k-1}^{\;\;i}V_R(x-1) + {}_{k-1}^{\;\;i}V_L(x+1)] \\ &\quad + \tau[({}_{k-1}\phi(m-1) + {}_{k-1}\phi(m+1)] \\ &= \Gamma\{\Gamma\,[{}_{k-2}^{\;\;i}V_R(x) + {}_{k-2}^{\;\;i}V_L(x))] + \tau_{k-2}\phi(m)\} \\ &\quad + \tau[{}_{k-1}\phi(m-1) + {}_{k-1}\phi(m+1)] \end{aligned} \tag{8.15}$$

(where $\Gamma = \rho - \tau$)

we can then substitute the earlier nodal voltage ${}_{k-2}\phi(x)$ which is given by

$$_{k-2}\phi(x) = {}_{k-2}^{\;\;i}V_L(x) + {}_{k-2}^{\;\;i}V_R(x) + I(x)\left[\frac{R+Z}{2}\right] \tag{8.16}$$

back into (8.15) to get:

$$\begin{aligned} {}_k^iV_R(x) + {}_k^iV_L(y) = \Gamma\left\{\Gamma\left[{}_{k-2}\phi(x) - I(x)\left[\frac{R+Z}{2}\right]\right] + \tau_{k-2}\phi(m)\right\} \\ + \tau[{}_{k-1}\phi(m-1) + {}_{k-1}\phi(m+1)] \end{aligned} \tag{8.17}$$

This in turn can be substituted into eqn. (8.14) to give ${}_k\phi(x)$:

$$\begin{aligned} &= [\Gamma^2 + 2\tau\Gamma]\,{}_{k-2}\phi(x) + \tau\,[{}_{k-1}\phi(m-1) + {}_{k-1}\phi(m+1)] \\ &\quad + [1-\Gamma^2]\,I(x)\left[\frac{R+Z}{2}\right] \end{aligned} \tag{8.18}$$

We can now move this to the new time and make the simplification that

$$[\Gamma^2 + 2\tau\Gamma] = \Gamma$$

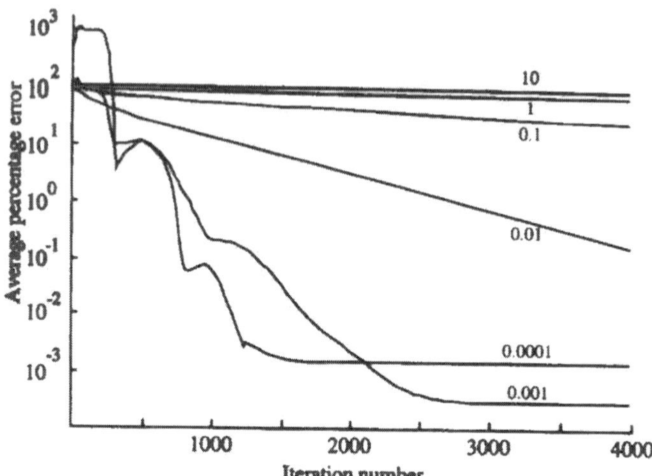

Figure 8.3 The convergence of the average percentage error in the TLM solution of the potential distribution in a GaAs diode as a function of damping resistor, R (note that $Z = 1$ so that $r = R/(1+R)$).

which gives ${}_{k+1}\phi(x)$:

$$= [\Gamma^2 + 2\tau\Gamma]_{k-1}\phi(x) + \tau[_k\phi(m-1) + {}_k\phi(m+1)] + [1-\Gamma^2]\, I(x)\left[\frac{R+Z}{2}\right] \tag{8.19}$$

Next, we take the Taylor expansion of both sides of this equation. After cancellation and setting $\Delta t = 1$ (the iteration time) we have:

$$\frac{\partial^2\phi}{\partial x^2} = \left[\frac{1}{\Delta x^2}\right]\frac{\partial^2\phi}{\partial t^2} + \left[\frac{1}{\Delta x^2}\right]\frac{\rho}{\tau}\frac{\partial\phi}{\partial t} - \left[\frac{2RI(x)}{\Delta x^2}\right] - \left[\frac{\Delta x^2}{12}\right]\frac{\partial^4\phi}{\partial x^4} \tag{8.20}$$

Thus, there is a spatial error $(\Delta x^2/12)\ (\partial^4\phi/\partial x^4)$
The dynamic components in this equation: $(1/\Delta x^2)\{(\partial^2\phi/\partial t^2) + (\rho/\tau)$ $(\partial\phi/\partial t)\}$ determine how fast the system converges and draws us back to some of the observations in the Laplace section on the importance of ρ/τ.

In order to examine these in a practical way Chakrabarti [8.2] took a model of a 4 µm gallium arsenide np junction diode for which voltage distribution results were available for a reverse bias of 2V. The problem space was divided into 500 nodes with the junction situated in the middle. An average percentage error was calculated in terms of the mean of the difference between the data for any point and the equivalent TLM calculated result. This is plotted against iteration number for different levels of damping and is shown in figure 8.3. There are a series of steps which correspond to the data at the junction undergoing amendment on reaching a fixed value boundary (~250 iterations) and the other boundary after one transit (~750 iterations) with repeats after subsequent transits of the mesh. It is also interesting to note that $R = 0.001$ gives a lower final error than does $R = 0.0001$ which is consistent with the observations in the case of Laplace equation.

References

[8.1] D. de Cogan, A. Chakrabarti and R.W. Harvey, TLM algorithms for Laplace and Poisson fields in semiconductor transport, *SPIE Proceedings*, **2373** (1996) 198–206.

[8.2] A. Chakrabarti, Transmission Line Matrix modelling for semiconductor transport, *PhD thesis*, UEA, Norwich 1996.

Chapter 9

TLM and inverse problems

Introduction

The very nature of the TLM scattering process seems to lend itself to the solution of inverse problems. Under normal operation it is a scattering process which runs forward in time in a deterministic way. So, the application of the same rules running in the opposite sense should be equivalent to running backwards in time. There are those who say that this is not possible since such processes are not inherently stable. They see the scattering in forward time as attenuation and the therefore the opposite process must be amplification. In the work that is reported here there have been no major stability problems.

It is true that it is an ill-posed problem and generally has more unknowns than equations available for solution. Nevertheless, it is possible to run a TLM algorithm in the reverse mode and by a process of optimisation identify some input or scattering event that has led to a particular set of data at the time of observation. In this chapter we will outline the various approaches that have been used, but will concentrate on those areas which are associated with thermal conduction processes.

TLM and imaging

The treatment of inverse problems are already well known for 'lossless' propagation and form the basis of imaging such as ultrasonic foetal monitoring, radar and sonar. Aspects of the application of TLM in these areas have been addressed in two papers in the past, but until now the subject has really not been brought to conclusion.

Dick Stevens while working with Peter Johns identified three headings which must be considered en-route to the development of an imaging scheme using TLM.

(1) How does the excitation signal behave in a material of known properties?
(2) What is the information content of the scattered wave?
(3) How does one use the changes in this information (resulting from the variation of propagation parameters) to determine values which would allow a model to reproduce the experimentally determined information?

The Johns and Stevens paper [9.1] paid most attention to the first two topics because: "... the third has been considered previously". By this they meant that there was already a body of work on impediography and several references were cited (e.g. Jones [9.2] and Mittra [9.3]). The first heading above is well known and has been the subject of the initial chapters of this book. We recall that if we were to propagate a pulse across a mesh we would observe serious dispersion effects. The authors discuss the implications of this and suggest filtering techniques which remove all frequencies above the Nyquist limit and significantly reduce the components within the TLM dispersion range.

At this stage they then construct a general propagation model where the local permittivities can be altered. The objective was to use optimisation so that the decimated (filtered) output at the observation points reproduce the experimentally observed sample sequence in a real medium whose properties are unknown. The assumption was that if the TLM sequence did match the measured values then the model would be an accurate representation of that medium.

Orme *et al* [9.4] addressed a slightly different type of problem. Peter Johns in private discussions with this author indicated that he hoped to apply TLM to the analysis of underwater acoustic scattering data for subsurface feature identification. The speed of sound in water creates very significant problems. The target and propagation distances are often large compared with the interrogating wavelength. Thus, if we are to have even 5 nodes per wavelength we will be faced with the processing of enormous meshes. On the other hand, ray tracing techniques are suitable for long-range propagation, but they do run into problems when local target features gives rise to diffractive scattering. A hybrid approach was proposed by Orme *et al* whereby the regions around both the acoustic source and target would be enmeshed and subject to TLM analysis, while ray propagation would be used for the intervening space. Two questions arise immediately:

(1) how do we describe the excitation of the TLM mesh by an incoming ray?
(2) how do we detect new wave-fronts which should be launched as rays from the mesh?

Their paper concentrates on these two issues and the extension to inverse analysis is not discussed in detail.

Inverse treatments of thermal problems

The situation is somewhat different in inverse thermal modelling and the goals have perhaps been less ambitious. For one thing mesh dispersion is not a problem. The information content of TLM signals which are observed in space and time have been subjected to heavy filtering on account of the low-pass nature of the nodes.

de Cogan and Soulos [9.5] have developed an algorithm which operates TLM in reverse in order to determine the position and magnitude of an excitation which was the source of the observed temperature profile. It can also be used to estimate the lapsed time between the initial excitation and

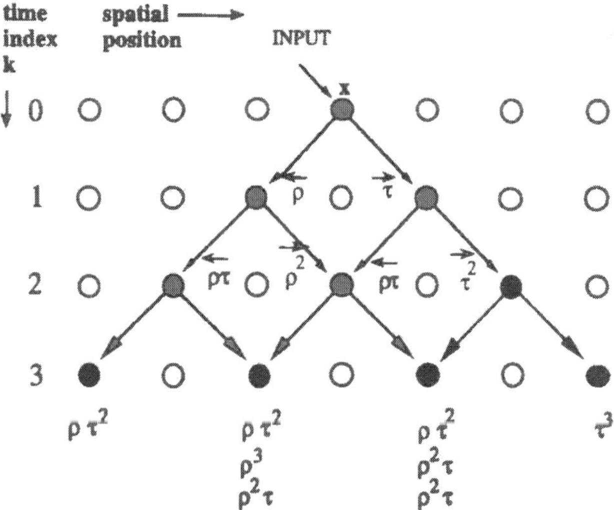

Figure 9.1 The response of a one-dimensional link-line TLM mesh during the first three iterations after a single unit excitation from the left.

measured profile. Until now, the treatment has not been extended beyond a single shot input.

Soulos [9.6] starts with a scattering analysis similar to that which was outlined in Chapter 4. The scatter diagram in figure 9.1 shows the events following the injection of a single pulse from the left onto node x. The instantaneous population of impulses on the nodes at iteration, $k = 3$ is shown immediately below the diagram.

If we look at the situation at nodes $x + 1$ and $x - 1$ at time $k = 3$ then we can write the relationships:

$$\begin{aligned} {}_3^iV_L(x+1) &= \rho\,{}_2^iV_R(x) + \tau\,{}_2^iV_L(x) \\ {}_3^iV_R(x-1) &= \rho\,{}_2^iV_L(x) + \tau\,{}_2^iV_R(x) \end{aligned} \tag{9.1}$$

Since we will only deal with incident pulses in this analysis we will develop some short-hand symbols. In the presentation which follows $3^i V_L(x + 1)$ will be written as ${}_3L(x + 1)$, ${}_3^iV_R(x - 1)$ as ${}_3R(x - 1)$ and so on. Thus, eqn (9.1) becomes:

$$\begin{aligned} {}_3L(x + 1) &= \rho\,{}_2R(x) + \tau\,{}_2L(x) \\ {}_3R(x - 1) &= \rho\,{}_2L(x) + \tau\,{}_2R(x) \end{aligned} \tag{9.2}$$

It should be clear that

$$\begin{aligned} {}_2R(x) &= \frac{{}_3L(x + 1) - \tau\,{}_2L(x)}{\rho} \\ {}_2L(0) &= \frac{\rho\,{}_3R(x-1) - \tau\,{}_3L(x+1)}{\rho^2 - \tau^2} \end{aligned} \tag{9.3}$$

This can be expressed in a general way as:

$$\begin{aligned} {}_{k-1}R(x) &= \frac{{}_kL(x + 1) - \tau\,{}_{k-1}L(x)}{\rho} \\ {}_{k-1}L(x) &= \frac{\rho\,{}_kR(x - 1) + \tau\,{}_3L(x + 1)}{\rho^2 - \tau^2} \end{aligned} \tag{9.4}$$

So long as $\rho \neq \tau$ or $\rho \neq 0$ then the application of the above analysis to the expressions at $k=3$ in figure 9.1 will result in a smaller occupied space at each reverse step. Ultimately, the minimum excited space reduces to one node which represents a unit excitation from the left three steps back.

Table 9.1

Δt	*Iterations*	*Estimate of input*	*Error (%)*
0.1	10	1301.2	+30.12
0.05	20	1136.9	+13.69
0.01	100	1000	0
0.005	200	981.9	−1.89
0.001	1000	967.3	−3.27

Assuming that we had a temperature profile arising from a single impulse (from the left) at some time in the past, then we would be faced with a problem: what is the most appropriate value of Δt? It should be remembered that Δt determines the values of ρ and τ through:

$$\rho = \frac{1}{1 + D\left[\frac{\Delta t}{\Delta x^2}\right]} \tag{9.5}$$

we could use a value of ρ corresponding to a large value of Δt. In this case it is possible that our back-stepping might be too coarse to accurately identify the source in space and time.

As an example we can consider the hypothetical case of a single pulse of magnitude 1000 injected (from the left) into a material whose thermal

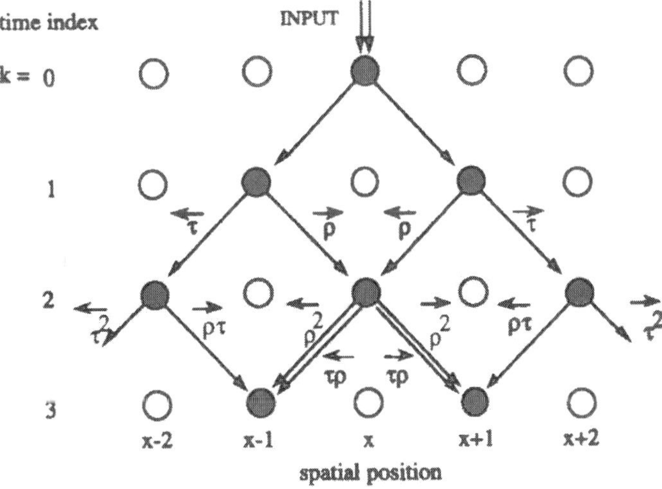

Figure 9.2 The response of a one-dimensional link-line TLM mesh to a real unit excitation during the first three iterations.

diffusivity was 1.17 cm^2/sec. A forward TLM routine run for 100 iterations with $\Delta x = 0.2$ cm and $\Delta t = 0.01$ sec provides a profile which is the 'experimental' data for an inverse algorithm. This algorithm does not assume any prior knowledge and therefore different values of ρ and τ must be used, corresponding to different guesses of the most appropriate value of Δt. The estimates of the input at minimum excited space and the resulting errors are shown in Table 9.1.

Further developments by de Cogan and Soulos [9.5] were based on realistic profiles which might result from a 'real' single excitation which initially divides into equal left and right-moving components (see figure 9.2).

Let us suppose that we have a single profile ($\mathsf{T}_1, \mathsf{T}_2, \mathsf{T}_3, \mathsf{T}_4, \mathsf{T}_5, \mathsf{T}_6$ and T_7) at a time which we will arbitrarily designate as $k=5$. A view of the immediate pre-history is shown in figure 9.3.

The scattered pulses which contribute to temperatures at $k = 5$ are labelled as A_1, A_2, A_3, B_1, B_2 and B_3, and it is immediately obvious that $A_1 = 0$ since it would arise from a point outside the $(x \pm k)$ region. A set of equations can then be constructed to express the temperatures at $k = 6$:

$$\begin{aligned}
\mathsf{T}_1 &= \mathsf{T}_7 = \tau B_1 \\
\mathsf{T}_2 &= \mathsf{T}_6 = \rho A_2 + \tau B_2 + \rho B_1 \\
\mathsf{T}_3 &= \mathsf{T}_5 = \rho A_3 + \tau B_3 + \rho B_2 + \tau A_2 \\
\mathsf{T}_4 &= \rho B_3 + \tau A_3 + \rho B_3 + \tau A_3
\end{aligned} \tag{9.6}$$

The sans-serif type-face indicates known or measured quantities. Thus we have a set of four equations with five unknowns. We can overcome this by treating one of the unknowns, e.g. A_3 as a variable in a matrix equation. The

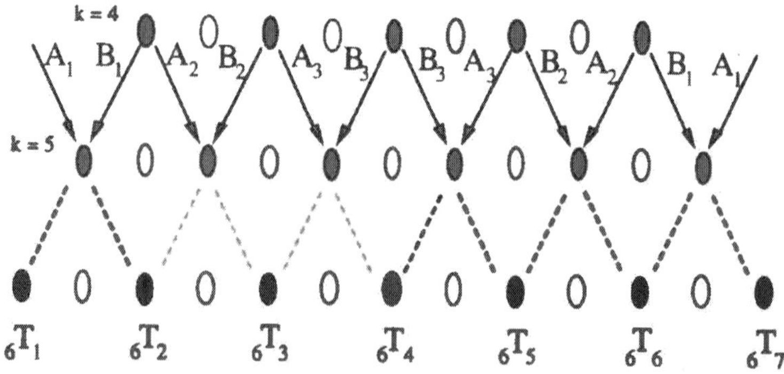

Figure 9.3 The immediate pre-history of a set of temperatures observed at iteration, $k=6$.

value of A_3 can then be identified by a simple search method.

$$\begin{bmatrix} \tau & 0 & 0 & 0 \\ \rho & \tau & 0 & \rho \\ 0 & \rho & \tau & \tau \\ 0 & 0 & \rho & 0 \end{bmatrix} \begin{bmatrix} B_1 \\ B_2 \\ B_3 \\ A_2 \end{bmatrix} = \begin{bmatrix} \mathbf{T}_1 \\ \mathbf{T}_2 \\ \mathbf{T}_3 - \rho A_3 \\ \dfrac{\mathbf{T}_4 - 2\tau A_3}{2} \end{bmatrix} \quad (9.7)$$

Choosing an arbitrary value for A_3 will always provide solutions for B_1, B_2, B_3 and A_2 but many of these will be negative. However, it was found that if one started with $A_3 = 0$ and progressed upwards, then the first value of A_3 which provided non-negative solutions for each of B_1, B_2, B_3 and A_2 was a close approximation to the true value. Knowing the material parameters and hence the values of ρ and τ, the process could then be run backwards in time until the initial location, magnitude and time of excitation is identified.

The case described above was for 6 steps in time between excitation and collection of data. If the routine had obviously not gone through a minimum excitation area, then it would be necessary to incorporate more steps. If this were done there would be an increase in the number of unknowns and therefore in the number of variables to be optimised. This is summarised in Table 9.2.

TNDT using TLM

A variant of the inverse technique has also been used [9.7] to identify subsurface inhomogeneities. However, the first step is to locate such features. Taken together, these constitute a form of 'thermal non-destructive testing' (TNDT). The method which was used is called 'two-sided inspection'. This

Table 9.2

No. of Iterations	*No. of equations*	*No. of unknowns*	*No of variables*
2	2	1	0
4	3	3	0
6	4	5	1
8	5	7	2
10	6	9	3
12	7	11	4
14	8	13	5
–	–	–	–
100	51	99	48

involves applying heat on one side of a sample and examining the thermal profile at the other. We can define 'contrast' as the difference between the profile in the sample-under-test and a perfectly homogeneous sample measured at the same time after the application of rear-surface heating. If observations are made early, then the contrast is low and there is little information. The situation is similar if observations are made late, *i.e.* when the system has almost reached equilibrium.

(1) Defect location

Vavilov and Taylor [9.8] have developed a theory which relates surface temperature distribution to the topology of any sub-surface inhomogeneity or defect. They recommend that measurements should be made at maximum contrast. They also showed that the spatial derivative of the contrast plot goes through a minimum/maximum at locations on the surface which correspond to edges of the defect. Soulos [9.6] demonstrated these points using TLM. He took an analogue sample which comprised a 50 node wide inspection surface. The depth was represented by 25 nodes. A constant temperature heat source (200°C) was attached to the lower face at $t = 0$. A TLM forward simulation was run and the results represented the behaviour of a perfectly homogeneous sample.

A single 14 × 5 node void ($\rho_{\text{defect}} = 1$) was then placed within this sample between $x = 18$ and $x = 32$ and between $z = 6$ and $z = 10$. The heating simulation was repeated and the temperature at every position on the inspected surface at any time was observed, $T'(x, 0, k\Delta t)$. The absolute value of the contrast as a function of position and iteration time is shown in

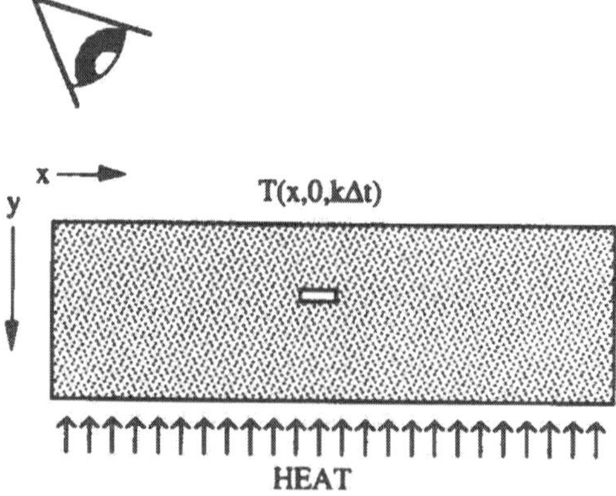

Figure 9.4 Double-sided inspection for defect location.

figure 9.5. The peak of this surface represents the point of maximum contrast in space and time, which confirms that surface inspection of temperature can be used to locate the position of a defect. This effect was also confirmed in cases where substantial non-void inhomogeneities had reflection coefficients which were either greater or smaller than that of the host material.

The second derivative of the contrast goes through maxima which correspond with the geometric edges of the void. It was found that this generally applied even when the void was located some distance below the inspection surface. It was also found that the same observations applied to non-void defects so long as the diffusivity was different from that of the host material.

The approach outlined above can only provide an indication of the dimensions of a defect as viewed in the $x-y$ plane. de Cogan *et al* [9.8] demonstrated that the time to maximum contrast provides a useful method for locating the depth of a defect below the inspection surface.

(2) Defect characterisation

The optimisation concepts which were used in inverse simulations were then extended to the characterisation of inhomogeneities. In this case it was necessary to introduce the diffusivity as an independent parameter into the

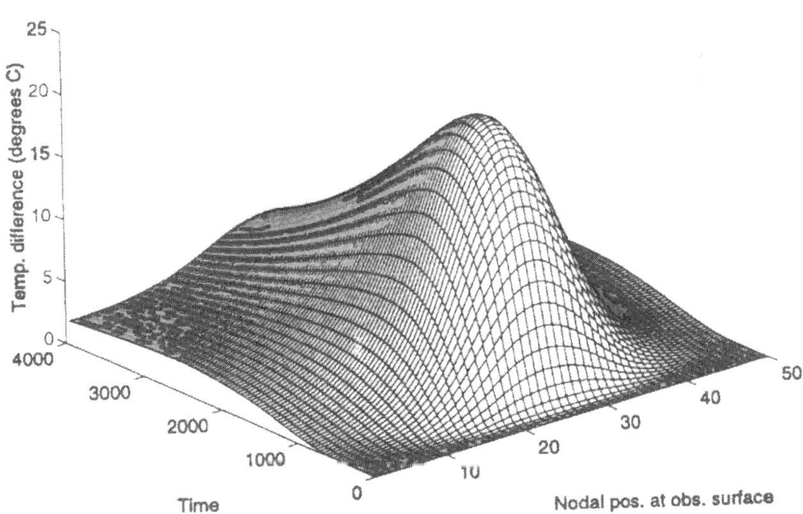

Figure 9.5 Contrast versus space and time for a void in an otherwise homogeneous sample subject to heating on the reverse side.

TLM equations. The TLM equations which describe the scattering process in three-dimensions contain a term, $Z/(Z+R)$ which needs to be represented in terms of diffusivity. In three-dimensions the equations for Z and R are:

$$Z = 3\,\frac{\Delta t}{C_p d \Delta x^3} \qquad R = \frac{1}{2k_T \Delta x} \tag{9.8}$$

where d is density, C_p specific heat and k_T is thermal conductivity.

Substituting these into $Z/(Z+R)$ gives:

$$\frac{6\Delta t}{(1/D)\Delta x^2 + 6\Delta t} \quad \left(\text{where} \quad D = \frac{k_T}{C_p d}\right) \tag{9.9}$$

If an inhomogeneity, has a thermal diffusivity that is different from the surrounding material then there will be an impedance discontinuity at any non-matching interface. Accordingly, another term must be included in the treatment. This is the interface reflection coefficient and is defined below for heat moving from the uniform specimen into a 'defect':

$$\frac{Z_{\text{defect}} - Z_{\text{specimen}}}{Z_{\text{defect}} + Z_{\text{specimen}}} = \frac{C_{p\,\text{specimen}}\, d_{\text{specimen}} - C_{p\,\text{defect}}\, d_{\text{defect}}}{C_{p\,\text{specimen}}\, d_{\text{specimen}} + C_{p\,\text{defect}}\, d_{\text{defect}}} \tag{9.10}$$

For heat transfer from 'defect' to specimen this is defined with opposite sign.

There are then two inhomogeneity parameters which must be optimised. The first is the diffusivity and the second is the product, $C_{p\,\text{defect}}\, d_{\text{defect}}$ which is required to describe the impedance mismatch between 'defect' and specimen.

In this work [9.7] information on the dimensions of the defect was used in conjunction with the temperature profile data to identify the material diffusivity. Additional information was also required. This included the time interval between excitation and collection of the data as well as the initial conditions of the experiment *i.e.* type of input and duration. The above data was then used in a series of 'blind' three-dimensional TLM simulations which provided modelled values of the inspected surface temperature profile. Diffusivity and thermal capacitance per unit volume were optimised until the results matched the 'experimental' data. The Levenberg-Marquardt method of optimisation [9.9] was found to work excellently for this type of problem and was used throughout.

References

[9.1] P.B. Johns and W. N. R. Stevens, Imaging by numerical analysis of wave scattering in the time-domain using transmission line modelling, signal processing and optimisation *IEE Proc. Part A*, **129**, (1982) 190 – 197.

[9.2] J.P. Jones, Impediography: a new ultrasonic diagnostic technique for diagnostic medicine, Ultrasound Med. and Biol., **1**, (1975) 489 – 497.

[9.3] R. Mittra, Inverse scattering and remote probing (chapter 7) in 'Computer techniques for Electromagnetics' (Ed R. Mittra), Pergamon Press 1973.

[9.4] E. A. Orme, P. B. Johns and J. M. Arnold, A hybrid modelling technique for underwater acoustic scattering, *International Journal of Numerical Modelling*, **1**, (1988) 189 – 206.

[9.5] D. de Cogan and A. Soulos, Inverse thermal modelling using TLM, *Numerical Heat Transfer* (Part B) **29**, (1996) 125 – 135.

[9.6] A. Soulos, TLM and inverse modelling in thermal diffusion problems, *MPhil thesis*, UEA, Norwich 1993.

[9.7] D. de Cogan, A. Soulos and K. O. Chichlowski, Sub-surface feature location and characterisation using inverse TLM techniques, *Proceedings 2nd Therminic Workshop*, Budapest, 25 – 27 September 1996, pp. 277 – 281.

[9.8] V. P. Vavilov and R. Taylor, Theoretical and practical aspects of the thermal non-destructive testing of bonded structures, *Research Techniques for Non-destructive Testing*, **5**, (1982) 239 – 279.

[9.9] W. H. Press, B.P. Flannery, S.A. Teukolsky and W.T. Vetterling, Numerical Recipes in C, Cambridge University Press, 1994 (section 15.5)

Chapter 10

Other Applications

This chapter outlines several topics which are important on account of the novelty of the application and which could not be comfortably included in earlier chapters.

TLM modelling of chemical kinetics

de Cogan and Henini [10. 1] developed a semi-empirical treatment which effectively modelled distributed loss on a transmission line. This was equivalent to first order chemical kinetics. The decay of reactant in the process:

$$A \xrightarrow{k} B$$

can be expressed as

$$-\frac{d[A]}{dt} = k[A] \tag{10.1}$$

the square brackets denote the concentration of A and k is the rate constant. If the initial concentration of reactant is $[A]_0$ then the concentration at time t is given by:

$$[A]_t = [A]_0 \, e^{-kt} \tag{10.2}$$

In terms of TLM, a pulse of magnitude ${}_k^sV_R(x-1)$ is considered as analogous to $[A]_0$. When it enters a transmission line of length Δx then it will take time Δt before it becomes incident at node x. Since the line is lossy there will be decay during transit so that we can write the incidence process as

$$\begin{aligned} {}_{k+1}^{i}V_L(x) &= {}_{k}^{s}V_R(x-1)\,e^{-k\Delta t} \\ {}_{k+1}^{i}V_R(x) &= {}_{k}^{s}V_L(x+1)\,e^{-k\Delta t} \end{aligned} \tag{10.3}$$

Thus it is possible to model a diffusion/recombination process in this way or equally to model a first order reaction of a diffusing chemical species.

We should be able to apply this to simple second order kinetics. For example, the decrease in the concentration of A in the reaction $2A \xrightarrow{k} B$ is:

$$-\frac{d[A]}{dt} = k[A]^2 \text{ (with concentration } [A]_0 \text{ at } t = 0) \tag{10.4}$$

Thus

$$[A]_t = \frac{[A]_0}{1+kt} \tag{10.5}$$

In TLM terms the connect process would then be:

$$\begin{aligned} {}_{k+1}^{i}V_L(x) &= \frac{{}_{k}^{s}V_R(x-1)}{1+k\Delta t} \\ {}_{k+1}^{i}V_R(x) &= \frac{{}_{k}^{s}V_L(x+1)}{1+k\Delta t} \end{aligned} \tag{10.6}$$

Problems arise when we try to apply the same ideas to reactions such as:

$$A + B \xrightarrow{k} C$$

The multi-compartment model due to Saleh [10.2, 10.3] offers a solution, but is restricted to spatially invariant concentrations. Nevertheless this is adequate for representing a well stirred reaction mixture. It encloses each species in an independent compartment which in electrical/electronic engineering terms is treated as an n-port network. The connections from the eight-port compartment in figure 1 denoted as g_1, g_2, g_3, g_4 or as d_1, d_2, d_3, d_4 represent the flow of species as a result of growth (g) and decay (d) to and from other compartments.

The topology of interconnection of compartments is merely a function of the reactions being modelled. The discrete components within a compartment (figure 10.2) consist of voltage and current generators as well as resistors which represent the kinetics relating to that species. The capacitor represents the storage of species within the compartment.

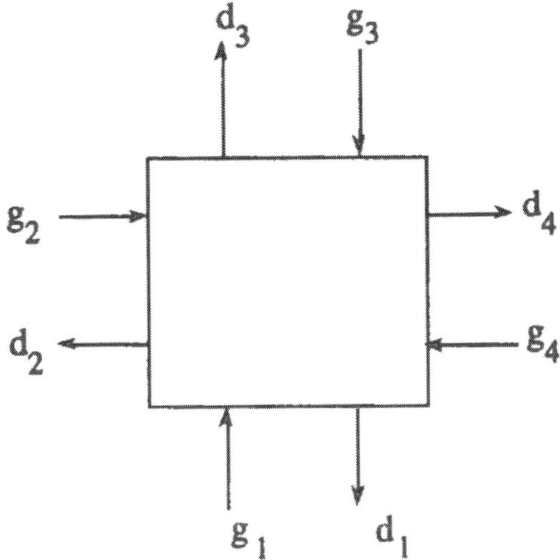

Figure 10.1 Single compartment with generation and decay.

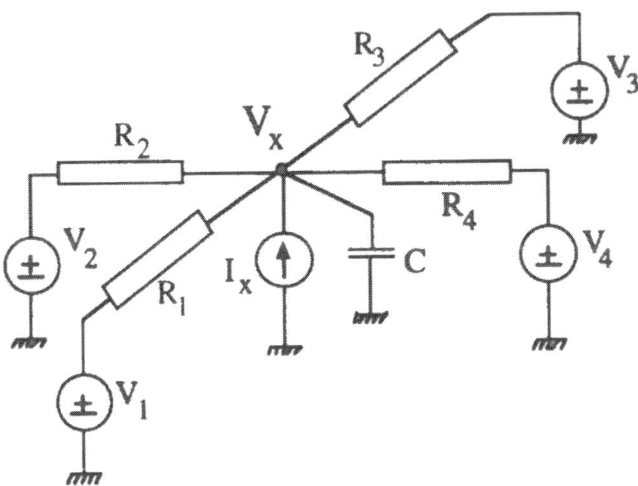

Figure 10.2 The lumped circuit components in a single compartment.

The transmission line equivalent circuit of a compartment is shown in figure 10.3. It is similar to figure 10.2 except for the open-circuit half-length stub, Z_r.

If a compartment is connected to four other compartments then the rate of change of concentration with time is given by:

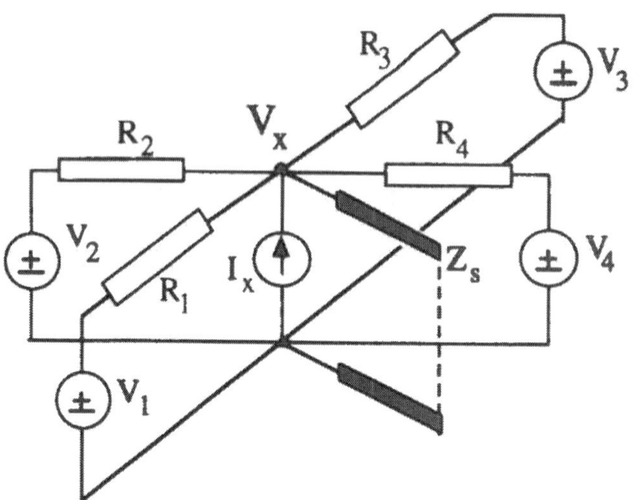

Figure 10.3 TLM equivalent of figure 10.2.

$$
\begin{aligned} d[X_0]/dt = & -(d_1 + d_2 + d_3 + d_4)[X_0] \\ & + g_1[X_1] + g_2[X_2] + g_3[X_3] + g_4[X_4] \end{aligned} \tag{10.7}
$$

d_1, d_2 etc. are the rates at which the reactant X_0 decays to species X_1, X_2 etc. as products. g_1, g_2 *etc.* are the rates at which product X_0 is generated from reactants X_1, X_2 etc.

A simple analysis of the circuit shown in figure 10.2 gives:

$$
C\frac{d[V_{X_0}]}{dt} = I_X - V_{X_0}\left[\frac{1}{R_1} + \frac{1}{R_2} + \cdots\right] + \frac{V_{X_1}}{R_1} + \frac{V_{X_2}}{R_2} + \cdots \tag{10.7}
$$

Assigning unit value to the capacitance C and setting $g_i = 1/R_i$

$$
\frac{d[V_{X_0}]}{dt} = I_X - V_{X_0}[g_1 + g_2 + \cdots] + V_{X_1}g_1 + V_{X_2}g_2 + \cdots \tag{10.8}
$$

The current generator is deliberately chosen so that $I_X = g_m V_X$ where

$$
g_m = \sum_1^4 (g_i - d_i) \tag{10.9}
$$

Thus $d[V_{X_0}]/dt$ is:

$$-(d_1 + d_2 + d_3 + d_4)[V_{X_0}] + g_1[V_{X_1}] + g_2[V_{X2}] + g_3[V_{X3}] + g_4[V_{X4}] \tag{10.10}$$

which is identical to the original rate equation.

Instead of the normal Z_s we will use the stub admittance, Y_s because it is easier to manipulate. It is chosen so that:

$$C = Y_s(\Delta t/2) \tag{10.11}$$

The superposition theorem gives a value of the voltage V_X:

$${}_kV_X = \frac{\sum_{j=1}^{4} ({}_kV_j\, g_j) + 2\,{}_k^iV_s\, Y_s + g_m\, {}_{k-1}V_X}{\sum_{j=1}^{4} (g_j + Y_s)} \tag{10.12}$$

The pulse scattered back into the stub ${}_k^sV_s$ is:

$${}_k^sV_s = {}_kV_X - {}_k^iV_s \tag{10.13}$$

This is reflected at the open circuit termination and becomes incident at the next iteration.

The initial values of the compartments are achieved by launching a voltage pulse into the appropriate stub at $t=0$. If A_0 is the initial concentration for compartment A, then the magnitude of the incident pulse is given by:

$${}_0^iV_s(A) = \frac{A_0 \sum_{j=1}^{4} (g_j + Y_s)}{2Y_s} \tag{10.14}$$

V_1, V_2, V_3, V_4, $V_X=0$ for all compartments for $t > 0$

Applications

A first order reaction demonstrates the translation between the chemical problem and the multi-compartment network.

$$A \xrightarrow{k_1} B$$

This is a two compartment problem with the compartments labelled a and b (the convention is to designate compartments by lower case letters and to denote the contents or concentration of species within a compartment by upper case letters)

In compartment 'a' all generation rates are zero as there is nothing replenishing the concentration of species A. In compartment 'b' the rate of generation of species B by the decay of species A, $g1(b) = k_1$. All other generation rates in this compartment are zero. There is only one line of decay: $d_1(a) = k_1$. All others in both compartments are zero. The terminal voltages in compartment 'a' are zero. $V_1(b) = V_X(a)$ (i.e. $V_1(b)$ for the next iteration assumes the recently calculated junction value in compartment 'a'). The other terminal voltages in 'b' are zero.

In general, a second order reaction can be treated similarly

$$2A \xrightarrow{k_1} B$$

We have that:

$$\frac{dV(a)}{dt} = -d_1(a)\,V_X(a) + g_1(a)\,V_1(a) \tag{10.15}$$

Therefore $d_1(a) = k_1 V_X(a)$, $g_1(a) = 0$

The rate of growth of concentration of species B is $d[B]/dt = k_1[A]^2$

This means that:

$$\frac{dV(b)}{dt} = -d_1(b)\,V_X(b) + g(b)\,V_1(b) \tag{10.16}$$

Therefore $d_1(b) = 0$, $g_1(b) = k_1$, $V_1(b) = V(a)^2$

All other parameters in any multi-port compartment are set to zero.

We can also develop the necessary equations for equilibrium reactions such as

$$A + B \underset{k_r}{\overset{k_f}{\rightleftharpoons}} C + D$$

The decay and growth of species can be summarised by the following equations:

$$\frac{d[A]}{dt} = -k_1[A][B] + k_2[C][D]$$
$$\frac{d[B]}{dt} = -k_1[A][B] + k_2[C][D]$$
$$\frac{d[C]}{dt} = -k_2[C][D] + k_1[A][B]$$
$$\frac{d[D]}{dt} = -k_2[C][D] + k_1[A][B]$$

The values of generation rates, decay rates and terminal voltages in the rate equations for the appropriate compartments are shown in Table 10.1.

Table 10.1

	a	*b*	*c*	*d*
d_1	$k_1V_x(b)$	$k_1V_x(a)$	$k_2V_x(d)$	$k_2V_x(c)$
d_2	0	0	0	0
d_3	0	0	0	0
g_1	k_2	k_2	k_1	k_1
g_2	0	0	0	0
g_3	0	0	0	0
V_1	$V_x(c)V_x(d)$	$V_x(c)V_x(d)$	$V_x(a)V_x(b)$	$V_x(a)V_x(b)$
V_2	0	0	0	0
V_3	0	0	0	0

A TLM simulation of the above equilibrium with an iteration time step, $\Delta t = 5 \times 10^{-4}$ sec was undertaken with the following normalised concentrations: $[A]_0 = 100$, $[B]_0 = 50$, $[C]_0 = 10$, $[D]_0 = 0$. The rate constants were: $k_f = 0.5$, $k_r = 0.02$. Figure 10.4 shows a plot of $[C(t)]\,[D(t)]\,/\,[A(t)]\,[B(t)]$ versus time and shows how equilibrium is approached for this set of concentrations.

The multi-compartment technique was developed in order to simulate the phenomenon of oscillating chemical reactions. At first this objective was not

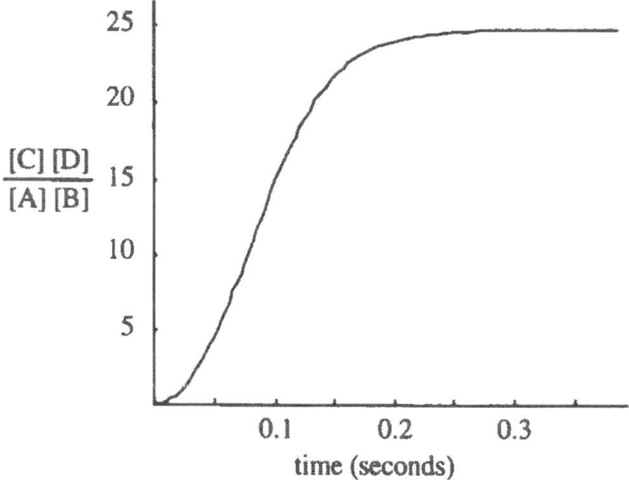

Figure 10.4 The TLM calculated convergence to equilibrium second order reactions described above.

achieved, although the technique was applied to the solution of a range on non-linear differential equations [10.4]. The source of this difficulty seems to have been in the definition of the current generator which is based on the voltage of the node at the previous time-step. Since equation (10.12) has no spatial dependence it is in fact possible to use the voltage at the present time:

$$_kV_X = \frac{\sum_{j=1}^{4} ({}_kV_j\, g_j) + 2\,{}_k^iV_s\, Y_s + g_m\, {}_kV_X}{\sum_{j=1}^{4} (g_j + Y_s)} \tag{10.17}$$

which reduces to:

$$_kV_X = \frac{\sum_{j=1}^{4} ({}_kV_j\, g_j) + 2\,{}_k^iV_s\, Y_s}{\sum_{j=1}^{4} (g_j + Y_s) - g_m} \tag{10.18}$$

This approach has been applied successfully to a solve a wide range of linear and non-linear differential equations. In particular, it has been applied to the Brusselator model for an oscillating chemical reaction system [10.5]. This considers the following reactions:

$$\begin{gathered} A \longrightarrow X \\ 2X + Y \longrightarrow 3X \\ B + X \longrightarrow Y + D \\ X \longrightarrow E \\ \text{equivalent to } A + B \longrightarrow E + D \end{gathered}$$

These can be reduced to the following differential equations:

$$\begin{aligned} \frac{d[X]}{dt} &= [A] + [X]^2[Y] - [B][X] - [X] \\ \frac{d[Y]}{dt} &= [B][X] - [X]^2[Y] \end{aligned} \tag{10.19}$$

From Glansdorff and Prigogine [10.6] we know that the system is in equilibrium when $[X]_0 = [A]$ and $[Y]_0 = [B]/[A]$ and that instability occurs if $[B] > 1 - [A]^2$. Beyond instability the steady-state enters a limit cycle.

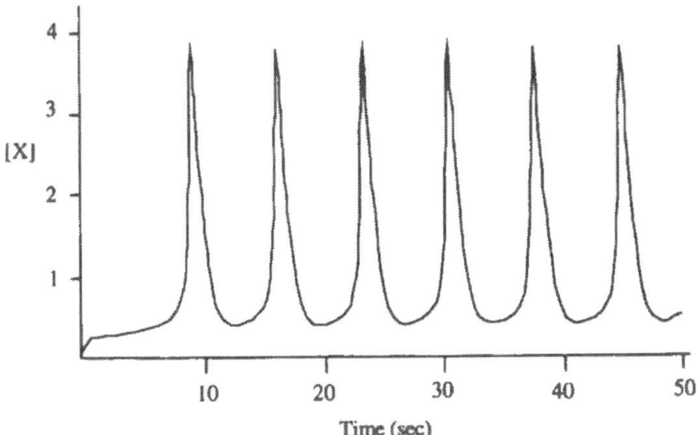

Figure 10.5 The TLM simulated concentration of intermediate X calculated using $[A]=1$, $[B]=3$ and $\Delta t=0.001$ sec.

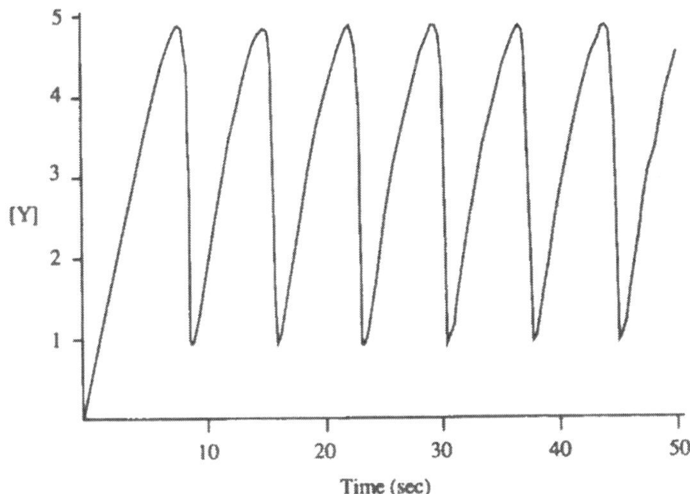

Figure 10.6 The TLM simulated concentration of intermediate Y calculated using $[A]=1$, $[B]=3$ and $\Delta t = 0.001$ sec.

The TLM simulation using equation (10.18) has been very successful. The correct limit-cycle behaviour has been observed. Figure 10.5 which shows the time variation of the concentration of intermediate X clearly demonstrates the oscillating nature of this type of reaction. Figure 10.6 shows the equivalent behaviour of intermediate Y. These results exhibit subtle differences when compared with those calculated using the Runge-Kutta method and this is currently under investigation. However, it should be noted that a comparison between results calculated using Runge-Kutta, Adams-Bashford and Adams-Moulton techniques also show differences [10.7].

Modelling of viscous flow

The vertical flow of a viscous fluid from an opening was characterised by Trouton [10.8]. A descending fluid (figure 10.7) at any distance y from the orifice has an area $A(y)$ and a velocity $v(y)$ and a rate of elongation $dv(y)/dy$.

The weight of material below y provides a tractive force such that:

$$\frac{F(y)}{A(y)} = \lambda \frac{dv(y)}{dy} \tag{10.20}$$

λ is the coefficient of viscous traction and is given by $\lambda = 3\eta$ (where η is coefficient of viscosity). We can start to develop a TLM model of this process by considering an elemental volume of fluid of mass $\rho A(y)dy$ so that:

$$\frac{dF}{dy} + \rho A g = \rho A \frac{dv}{dt} \tag{10.21}$$

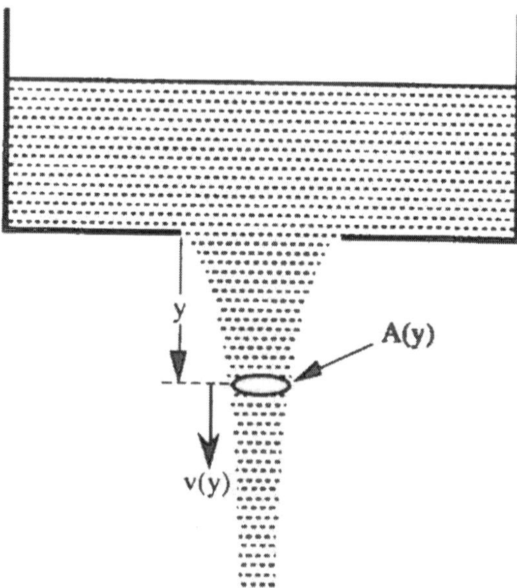

Figure 10.7 Viscous flow from an orifice.

If this substituted into the equation above we obtain

$$\frac{d^2v}{dy^2} + \frac{\rho}{\lambda} g = \frac{\rho}{\lambda}\frac{dv}{dt} \quad \left(\text{where } \frac{dv}{dt} = \frac{\partial v}{\partial t} + v\frac{\partial v}{\partial t}\right) \tag{10.22}$$

which is consistent with the Navier-Stokes equation. The problem can be described as diffusion of the effect of gravity causing relative motion within the body, the exact amount depending on the density and viscosity. This last equation could be rewritten as:

$$\frac{d^2v}{dy^2} + \frac{\rho}{\lambda}\frac{d(gt)}{dt} = \frac{\rho}{\lambda}\frac{dv}{dt} \tag{10.23}$$

The TLM equivalent for this would be:

$$\frac{d^2\Phi}{dy^2} + 2R_d\,C_d\,\frac{d(\Phi')}{dt} = 2R_d\,C_d\,\frac{\rho}{\lambda}\frac{d\Phi}{dt} \tag{10.24}$$

(where $\Phi' = g't$, where g' is constant)

This is identical to any TLM diffusion model with a term $[2R_d\,C_d\,d(\Phi')/dt]$ which could be considered as driving the equivalent circuit.

In a practical implementation a contribution $g'\Delta t$ is made at each time-step in the form of $g'\Delta t/2$ added to each incident pulse. Φ is the velocity, $\Delta x\,C_d$ is the elemental mass and the elemental resistance is given by:

$$R = \frac{\Delta x}{2\lambda A(y)} \tag{10.25}$$

The current is given by $I = C\,d\Phi/dt$ so that the net current between nodes is the viscous tractive force developed over a time period. For high values of λ it is valid to treat gravity as a current source of magnitude C_g.

Newton [10.9] has successfully used this approach to simulate the growth of fluid columns as part of research into the application of TLM to firing processes during the manufacture of vitreous china ware.

Image processing

As computing power has increased the subject of image processing has gained in prominence. Computer stored images may be manipulated for the purpose of recognition, classification or encoding. In many instances the amount of information may need to be reduced for these purposes. Images may also contain noise or may be blurred and enhancement may be the objective of processing in these cases. Within the topic of data reduction different images may have different criteria. In one circumstance it may be necessary to retain edge data, while in another application, area information may be the important factor. 'Diffusion' processing of images is already well established [10.10, 10.11], although it must be recognised that conventional diffusion will by its nature loose edges. The application of TLM in this area is relatively new.

Wilkinson *et al.* [10.12] drew attention to the analogy between the TLM equations and the discrete state-space equations. This analogy formed the basis of a view of the theory of lossless TLM which was discussed in chapter 4. Wilkinson and Pulko [10.13] have used image filtering as a further demonstration of the TLM-discrete state-space analogy.

In a one-dimensional example

$$\begin{aligned} x(t+1) &= f(x(t), u(t), t) = A\,x(t) + Bu(t) \\ y(t) &= g(x(t), u(t), t) = Cx(t) + Du(t) \end{aligned} \tag{10.26}$$

Description of the system is then a matter of identifying an appropriate vector of state variables, x and determining the functions f and g corresponding to the dynamic behaviour of the system.

In a system where the super-position principle holds we can write:

$$\begin{aligned} f(x(t), u(t), t) &= \mathbf{A}\,x(t) + \mathbf{B}\,u(t) = \mathbf{Tx}\,x(t) \\ g(x(t), u(t), t) &= \mathbf{C}\,x(t) + \mathbf{D}\,u(t) = y(t) \end{aligned} \tag{10.27}$$

$u(t)$ is the input, $y(t)$ is the output, $\mathbf{T}$ is the forward time-shift operator, $\mathbf{x}$ is the forward space-shift operator in one-dimension.

The TLM nodal output is

$${}_k y(x) = [c_1, c_2]\begin{bmatrix} {}_k^i V_1(x) \\ {}_k^i V_2(x) \end{bmatrix} \tag{10.28}$$

where c_1 and c_2 are the coefficients of the output matrix C in equation (4.56)

The nodal output can be related to the nodal input via a transfer function, G

$${}_k y(x) = G_k\,u(x) \tag{10.29}$$

The expression for the transfer function is derived in chapter 4:

$$G = [c_1, c_2]\left[\begin{bmatrix} \mathbf{T}\mathbf{x} & 0 \\ 0 & \mathbf{T}\bar{\mathbf{x}} \end{bmatrix} - \begin{bmatrix} \tau & \rho \\ \rho & \tau \end{bmatrix}\right]^{-1} \mathbf{B} \tag{10.30}$$

where the input matrix $\mathbf{B} = \begin{bmatrix} b_1 \\ b_2 \end{bmatrix}$

There is a similar derivation for the two-dimensional transfer function. The authors then discuss the effect of the relative magnitudes of R and Z on the rate and extent of spatial interaction. They point out that with a low value or R information propagates through the TLM equivalent network with little damping so that edge information is lost. In two-dimensions the elements of **B** and **C** (the output matrix) allows independent control of the filtering characteristics in the x and y directions.

·Harvey and Kenny [10.14] have started from the opposite side and have developed non-linear TLM diffusion algorithms as a range of image processing tools. They recognised that normal diffusion leads to the blurring of edges and therefore they have adopted the anisotropic diffusion equation:

$$\begin{aligned} \frac{\partial\phi(x, y, t)}{\partial t} &= \nabla\cdot(D(x, y, t)\nabla\phi(x, y, t)) \\ &= D(x, y, t)\nabla^2\phi(x, y, t) + \nabla D(x, y, t)\cdot\nabla\phi(x, y, t)) \end{aligned} \tag{10.31}$$

This allows edges to be preserved by arranging that diffusion is minimised in those regions where the gradient of intensity is large. The diffusion 'constant', D is normally taken as some non-linear function of the local gradient:

$$D(x, y, t) = g\,|\nabla\phi(x, y, t)|^2 \tag{10.32}$$

g is a non-linear factor and some popular choices include:

$$g(\xi) = e^{-\chi^2} \tag{10.33a}$$

and

$$g(\xi) = \frac{1}{1+\chi^2} \tag{10.33b}$$

(where $\chi = \xi/K$)

The problem has been solved for an image using a two-dimensional TLM node with stubs based on the generalised approach which was mentioned in chapter 4:

The nodal summation of pulses is standard for a two-dimensional mesh with unit line impedance and a stub consisting of R_s and Z_s:

$$ {}_{k+1}\phi(x,\, y) = \left[\frac{2(R_s + Z_s)}{1 + R + 4(R_s + Z_s)}\right] \sum_{j=1}^{4} {}_{k+1}^{\,i}V_j(x,\, y) + \left[\frac{2(1 + R)}{1 + R + 4(R_s + Z_s)}\right] {}_{k+1}^{\,i}V_j(x,\, y) \tag{10.34} $$

The scattered pulses are given by:

$$ {}_{k}^{s}V_j(x,\, y) = g_j(x,\, y)\left[\frac{5\lambda}{1 + R}\right] k\,\phi(x,\, y) + \left[\frac{R - 1}{R + 1}\right] {}_{k}^{i}V_j(x,\, y) \quad (j = 1..4) \tag{10.35} $$

$$ {}_{k}^{s}V_5(x,\, y) = \left[1 - \lambda \sum_{j=1}^{4} g_j(x,\, y)\right]\left[\frac{5Z_s}{R_s + Z_s}\right] {}_{k}\phi(x,\, y) + \left[\frac{R - 1}{R + 1}\right] {}_{k}^{i}V_5(x,\, y) \tag{10.36} $$

$j = 1, 2, 3, 4$ represent the directions of incidence and scatter along the lines (N, S, E and W used elsewhere) and $j = 5$ represents incidence and scatter in the stub. λ is the discretisation parameter which in this case is 1/4. The connection equations are as usual, remembering that in an open-circuit stub-line ${}_{k+1}^{\,i}V_5 = {}_{k}^{s}V_5$.

In their demonstration of this technique Harvey and Kenny have arranged for Z, R_s and Z_s to have unity values and the non-linearity factor, g has been chosen dependent on the first difference:

for instance on line 1 $\quad g_1(x,\, y) = g(\,|\,\phi(x,\, y) - \phi(x - 1,\, y)\,|\,)$

The constant K in (10.33*a* and *b*) allows the modeller to have a trade-off between diffusion time and structure preservation.

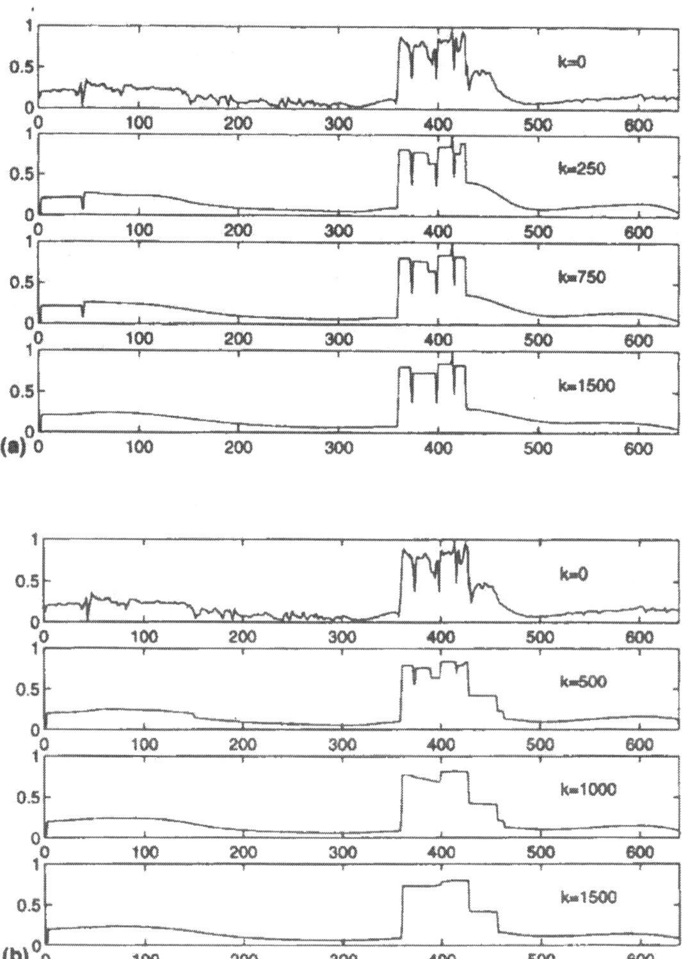

Figure 10.8 The intensity along a single image line as a function of the number of iterations for non-linear diffusion using eqns (10.33a and 10.33b).

Figure 10.8 shows the intensities of a single line of an image which have been treated using the variants of this algorithm. In figure 10.8a the edges are preserved due to the use of equation (10.33a), while the regions are preserved in figure 10.8b where equation (10.33b) was used. The preservation of edges as a function of the number of iterations is shown for a complete image in figure 10.9. Even after 1500 iterations it is clear that

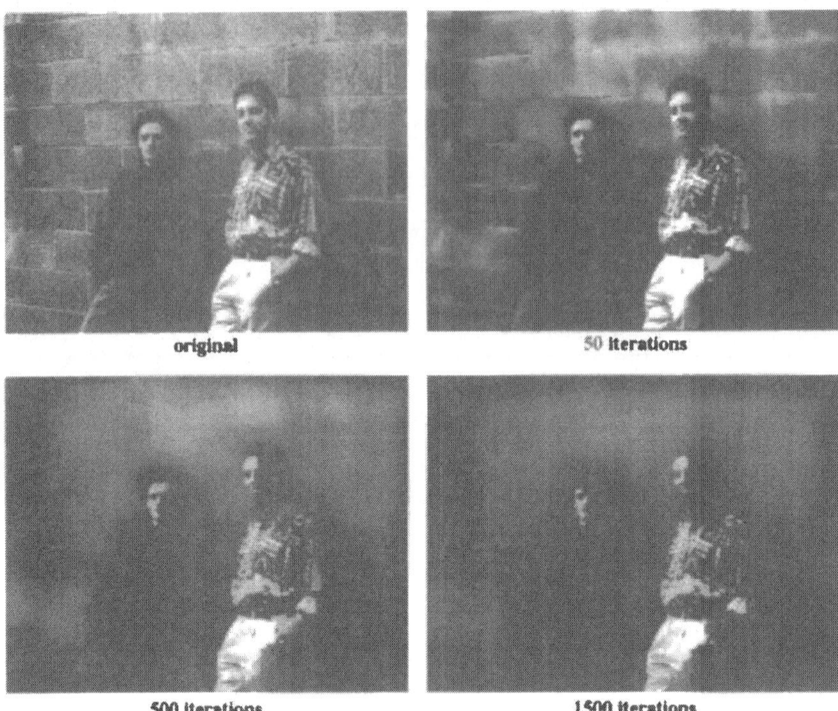

Figure 10.9 The effect of edge-preserving diffusion on an image as a function of the number of iterations ($K = 0.04$).

the details of what were the strongest edges in the original image are still retained.

The effect of a region preserving diffusivity (eqn (10.33b) as a function of number of iterations can be seen in the images in figure 10.10.

Wave mechanics

A wide range of numerical methods have been developed to solve Schrödinger's wave equation [10. 15] and it was not unnatural that attempts should be made to apply TLM to problems in this area. The subject gained its inspiration from the works of Kron who in 1945 published a paper on the use of an AC-network analyser for the Schrödinger equation [10. 16].

In 1995 Enders and de Cogan [10. 17] presented initial results on the application of TLM to the time-dependent Schrödinger equation. This was probably the first example of the use of a complex-valued transmission line

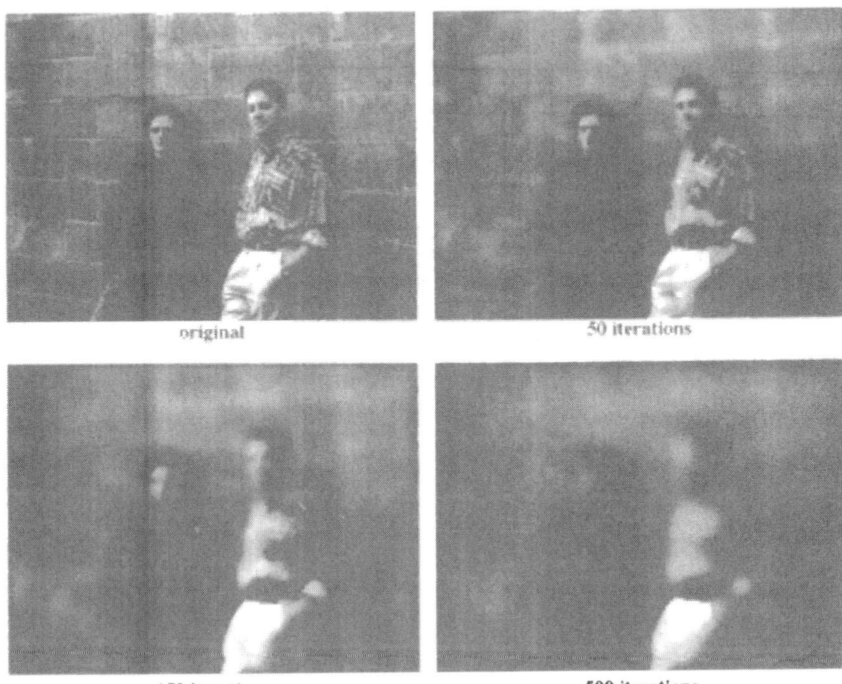

Figure 10.10 The effect of region-preserving diffusion on an image as a function of the number of iterations ($K = 0.04$).

network

$$j\hbar \frac{\partial \psi(x,\, t)}{\partial t} = -\frac{\hbar^2}{2m} \frac{\partial^2 \psi(x,\, t)}{\partial t^2} + V(x)\, \psi(x,\, t) \tag{10.37}$$

where $V(x)$ is a real valued function.

This is very similar to the diffusion equation with a recombination term except that the imaginary coefficient of the time derivative is connected with wave solutions. We can define a probability conservation law:

$$\begin{aligned} j\hbar \frac{\partial}{\partial t} \int_{-\infty}^{\infty} |\, \psi(x,\, t)\,|^2 \, dx &= -\frac{\hbar^2}{2m} \\ \int_{-\infty}^{\infty} \left[\psi^*(x,\, t) \frac{\partial^2 \psi(x,\, t)}{\partial t^2} + \psi(x,\, t) \frac{\partial^2 \psi^*(x,\, t)}{\partial t^2} \right] dx &= 0 \end{aligned} \tag{10.38}$$

where $\psi^*(x,t)$ is the complex conjugate.

Being quadratic in $\psi(x, t)$ this is essentially different from the linear mass conservation law of the diffusion equation:

$$\begin{aligned} \frac{\partial}{\partial t}\int_{-\infty}^{\infty} c(x,t)\,dx &= \int_{-\infty}^{\infty}\left[\frac{\partial^2 c(x,t)}{\partial t^2} - A\,c(x,t)\right]dx \\ &= -A\int_{-\infty}^{\infty} c(x,t)\,dx \end{aligned} \tag{10.39}$$

In the context of TLM this equation is concerned with charge conservation on the network.
The time dependent equation can be approached by adapting the conventional TLM algorithm for the network shown in figure 10.11. using $G_s = 1/R_s$ we can write:

$$_k\phi(x) = \frac{1}{1+(R+Z)(G_s/2)}\left[{}_k^iV_L(x) + {}_k^iV_R(x)\right] \tag{10.40}$$

$$\begin{aligned} {}_k^rV_L(x) &= \frac{R-Z}{R+Z}\,{}_k^iV_L(x) + \frac{Z}{R+Z}\,{}_k\phi(x) \\ {}_k^rV_R(x) &= \frac{R-Z}{R+Z}\,{}_k^iV_R(x) + \frac{Z}{R+Z}\,{}_k\phi(x) \end{aligned} \tag{10.41}$$

$$\begin{aligned} {}_{k+1}^{\;\;i}V_L(x) &= {}_k^rV_R(x-1) \\ {}_{k+1}^{\;\;i}V_R(x) &= {}_k^rV_L(x+1) \end{aligned} \tag{10.42}$$

Drawing parallels with diffusion/recombination the network analogues are then:

$$\frac{R}{Z} = \frac{\Delta x^2}{D\Delta t} = -j\,\frac{2m}{\hbar}\frac{\Delta x^2}{D\Delta t} \tag{10.43}$$

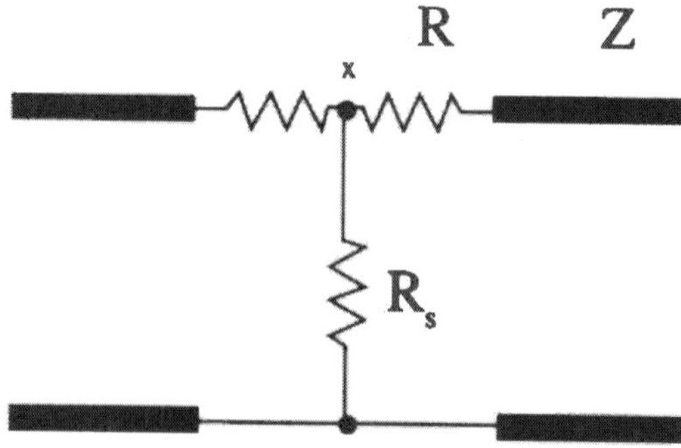

Figure 10.11 The one dimensional diffusion mesh with shunt loss used to model the Schrödinger equation.

$$G_s Z = A\Delta t = j\frac{V(x)\Delta t}{\hbar} \tag{10.44}$$

Thus if R and Z are held constant then G_s can be varied in proportion to $V(x)$. Both the damping resistor (R) and the leakage resistor (R_s) are imaginary valued which is consistent with the wave nature of equation. (10.37).

We can eliminate the reflected and nodal voltages:

$${}_{k+1}\begin{pmatrix} {}^iV_L \\ {}^iV_R \end{pmatrix} = \mathbf{P}\,{}_{k}\begin{pmatrix} {}^iV_L \\ {}^iV_R \end{pmatrix} \tag{10.45}$$

where

$$\mathbf{P} = \begin{pmatrix} \tau\bar{\mathbf{x}} & \rho\bar{\mathbf{x}} \\ \rho\mathbf{x} & \mathbf{x} \end{pmatrix}$$

$\mathbf{x}$ and $\bar{\mathbf{x}}$ are nodal shift operators such that $\mathbf{x}V(i) = V(i+1)$ and $\bar{\mathbf{x}}V(i) = V(i-1)$

The scattering parameters are given by:

$$\tau = \frac{1}{1+(R+Z)(G_s/2)}\left[\frac{Z}{R+Z}\right] \quad \text{and} \quad \rho = \frac{R-Z}{R+Z} + \tau \tag{10.46}$$

All of this mirrors the diffusion/recombination equation.

It was also shown that the TLM algorithm for the time-dependent Schrödinger equation is unconditionally stable. Initial tests using this algorithm to model the harmonic oscillator confirmed that Gaussian form of the initial approximation was preserved but there seemed to be considerable phase error. This has been the subject of further comment by Enders [5.18] who questions whether the current approach is capable of being taken further. The alternatives such as link-resistor node have yet to be investigated.

References

[10.1] D. de Cogan and M. Henini, Transmission line Matrix (TLM): A novel technique for modelling reaction kinetics, *J. Chem. Soc. Faraday Transactions*, **2**, 83 (1987) 843–855.

[10.2] A. H. M. Saleh, A multicompartment TLM model for solving first order rate equations, *Int. Jnl. Numerical Modelling*, **2** (1989) 53–60.

[10.3] A. H. M. Saleh and D. de Cogan, Numerical solution of inhomogeneous and non-linear differential equations using the TLM multi-compartment model, *Int. Jnl. of Numerical Modelling*, **3**, (1990) 215–228.

[10.4] A. H. M. Saleh and D. de Cogan, TLM models for limit cycle and other non-linear differential equations, *Proc. 4th Int. Conf. Non-linear Engineering Computations* (NEC-91), Sept. 1991, Pineridge Press, Swansea.

[10.5] J. J. Tyson, *J. Chem Phys*, 58 (1973) 3919.

[10.6] Glansdorff and Prigogine, *Thermodynamic theory of structure, stability and fluctuations*, John Wiley and Sons, New York 1971

[10.7] D. de Cogan and J. McClean, TLM modelling of oscillating chemical reactions (to be published).

[10.8] F. T. Trouton, On the coefficient of viscous traction and its relation to that of viscosity, *Proc. Royal Soc.*, **77** (1906) 426 – 440.

[10.9] H. Newton, TLM models of deformation and their application to vitreous china ware during firing, *PhD thesis*, Hull University 1994.

[10.10] T. Lindeberg and B. M. ter Haar Romeny Linear scale-space theory *in 'Geometry-driven diffusion in computer vision' (Ed. B. M. ter Haar Romeny) Kluwer Academic Publishers, Dordrecht 1994, 1* 38.

[10.11] P. Perona and J. Malik, Scale-space and edge detection using anisotropic diffusion, *IEEE Trans. Patt. Anal. Mach. Intelligence*, 12 (1990) 629 639.

[10.12] A. J. Wilkinson, S. H. Pulko and A. R. M. Witwit, A discrete state-space view of TLM, *Proc 9th IASTED-MIC*, (1990) 453 456.

[10.13] A. J. Wilkinson and S. H. Pulko, Image processing-an example of the TLM-State Space analogy, *Proceedings of MIC-92* (1992) 193 196 (ISBN 3-7153-0002-7).

[10.14] R. W. Harvey and C. P Kenny, *TLM for vision systems* First International Workshop on Transmission Line Matrix (TLM) Modelling-Theory and Applications, Victoria (BC) August 1995, 187 – 190.

[10.15] R. E. Mickens, Stable explicit schemes for equations of Schrödinger type, *Physical Review*, A 39 (1989) 5508 5510.

[10.16] G. K. Carter and G. Kron, AC network analyser study of the Schrödinger equation, *Physical Review*, 67 (1945) 44 - 49.

[10.17] P. Enders and D. de Cogan, TLM routines for the paraxial wave equation and the time-dependent Schrödinger equation, *Proc. First International Workshop on Transmission Line matrix (TLM) Modelling Theory and Applications*, Victoria BC, August 1995.

[10.18] P. Enders, Can the transmission line matrix method yield stationary solutions of the time-dependent Schrdinger equation?, *Int. Jnl. of Numerical Modelling*, 9 (1996) 321 – 323.

INDEX

An environmentally friendly book printed and bound in England by www.printondemand-worldwide.com

This book is made entirely of sustainable materials; FSC paper for the cover and PEFC paper for the text pages.

#0040 - 030114 - C0 - 229/152/12 [14] - CB